Lecture Notes in Physics

W0245748

Springer-Verlag Berlin Heidelberg GmbH

The Editorial Policy for Proceedings

The series Lecture Notes in Physics reports new developments in physical research and teaching – quickly, informally, and at a high level. The proceedings to be considered for publication in this series should be limited to only a few areas of research, and these should be closely related to each other. The contributions should be of a high standard and should avoid lengthy redraftings of papers already published or about to be published elsewhere. As a whole, the proceedings should aim for a balanced presentation of the theme of the conference including a description of the techniques used and enough motivation for a broad readership. It should not be assumed that the published proceedings must reflect the conference in its entirety. (A listing or abstracts of papers presented at the meeting but not included in the proceedings could be added as an appendix.)

When applying for publication in the series Lecture Notes in Physics the volume's editor(s) should submit sufficient material to enable the series editors and their referees to make a fairly accurate evaluation (e.g. a complete list of speakers and titles of papers to be presented and abstracts). If, based on this information, the proceedings are (tentatively) accepted, the volume's editor(s), whose name(s) will appear on the title pages, should select the papers suitable for publication and have them refereed (as for a journal) when appropriate. As a rule discussions will not be accepted. The series editors and Springer-Verlag will normally not interfere with the detailed editing except in fairly obvious cases or on technical matters.

Final acceptance is expressed by the series editor in charge, in consultation with Springer-Verlag only after receiving the complete manuscript. It might help to send a copy of the authors' manuscripts in advance to the editor in charge to discuss possible revisions with him. As a general rule, the series editor will confirm his tentative acceptance if the final manuscript corresponds to the original concept discussed, if the quality of the contribution meets the requirements of the series, and if the final size of the manuscript does not greatly exceed the number of pages originally agreed upon. The manuscript should be forwarded to Springer-Verlag shortly after the meeting. In cases of extreme delay (more than six months after the conference) the series editors will check once more the timeliness of the papers. Therefore, the volume's editor(s) should establish strict deadlines, or collect the articles during the conference and have them revised on the spot. If a delay is unavoidable, one should encourage the authors to update their contributions if appropriate. The editors of proceedings are strongly advised to inform contributors about these points at an early stage.

The final manuscript should contain a table of contents and an informative introduction accessible also to readers not particularly familiar with the topic of the conference. The contributions should be in English. The volume's editor(s) should check the contributions for the correct use of language. At Springer-Verlag only the prefaces will be checked by a copy-editor for language and style. Grave linguistic or technical shortcomings may lead to the rejection of contributions by the series editors. A conference report should not exceed a total of 500 pages. Keeping the size within this bound should be achieved by a stricter selection of articles and not by imposing an upper limit to the length of the individual papers. Editors receive jointly 30 complimentary copies of their book. They are entitled to purchase further copies of their book at a reduced rate. As a rule no reprints of individual contributions can be supplied. No royalty is paid on Lecture Notes in Physics volumes. Commitment to publish is made by letter of interest rather than by signing a formal contract. Springer-Verlag secures the copyright for each volume.

The Production Process

The books are hardbound, and the publisher will select quality paper appropriate to the needs of the author(s). Publication time is about ten weeks. More than twenty years of experience guarantee authors the best possible service. To reach the goal of rapid publication at a low price the technique of photographic reproduction from a camera-ready manuscript was chosen. This process shifts the main responsibility for the technical quality considerably from the publisher to the authors. We therefore urge all authors and editors of proceedings to observe very carefully the essentials for the preparation of camera-ready manuscripts, which we will supply on request. This applies especially to the quality of figures and halftones submitted for publication. In addition, it might be useful to look at some of the volumes already published. As a special service, we offer free of charge LATEX and TEX macro packages to format the text according to Springer-Verlag's quality requirements. We strongly recommend that you make use of this offer, since the result will be a book of considerably improved technical quality. To avoid mistakes and time-consuming correspondence during the production period the conference editors should request special instructions from the publisher well before the beginning of the conference. Manuscripts not meeting the technical standard of the series will have to be returned for improvement.

For further information please contact Springer-Verlag, Physics Editorial Department II, Tiergartenstrasse 17, D-69121 Heidelberg, Germany

Gérard Trottet (Ed.)

Coronal Physics from Radio and Space Observations

Proceedings of the CESRA Workshop
Held in Nouan le Fuzelier, France,
3–7 June 1996

 Springer

Editor

Gérard Trottet
Département d'Astronomie Solaire et Planétaire
Observatoire de Paris, Section de Meudon
5, Place Jules Janssen
F-92195 Meudon Principal Cedex, France

Frontispiece: Images of the solar corona obtained at 164 MHz by the Nançay Radioheliograph before and during the October 12, 1996 eclipse. The white circle indicates the visible solar disk and north is at the top. The image at 1250 UT was taken before the beginning of the eclipse (~1307 UT). It shows the typical appearance of the quiet corona at metric wavelengths with the presence of elongated coronal holes which extend southward from the northern polar regions. The other three images were taken during the eclipse. Maximum occultation (57% of the visible disk) occurred at ~1426 UT. The eclipse ended at ~1538 UT.

Cataloging-in-Publication Data applied for.

Die Deutsche Bibliothek - CIP-Einheitsaufnahme

Coronal physics from radio and space observations : proceedings of the CESRA workshop held at Nouan le Fuzelier, France, 3 - 7 June 1996 / Gérard Trottet (ed.).
(Lecture notes in physics ; Vol. 483)
ISBN 978-3-662-14120-5   ISBN 978-3-540-68693-4 (eBook)
DOI 10.1007/978-3-540-68693-4

ISSN 0075-8450
ISBN 978-3-662-14120-5

© Springer-Verlag Berlin Heidelberg 1997
Originally published by Springer-Verlag Berlin Heidelberg New York in 1997
Softcover reprint of the hardcover 1st edition 1997

The use of general descriptive names, registered names, trademarks, etc. in this publication does not imply, even in the absence of a specific statement, that such names are exempt from the relevant protective laws and regulations and therefore free for general use.

Typesetting: Camera-ready by the authors/editor
Cover design: *design & production* GmbH, Heidelberg
SPIN: 10550683   55/3144-543210 - Printed on acid-free paper

Nançay Radioheliograph
October 12, 1996 Solar Eclipse

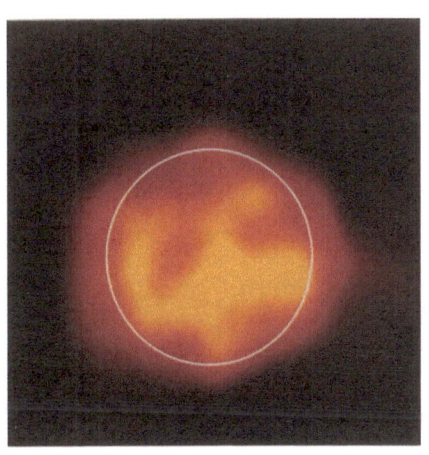

12 50 UT

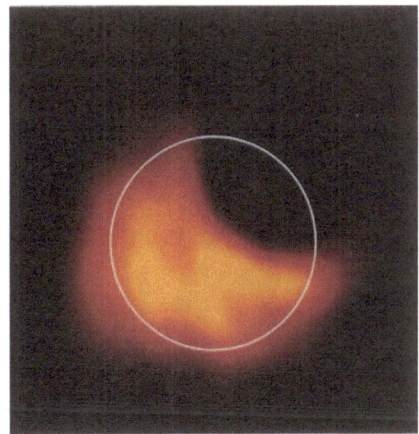

13 51 UT

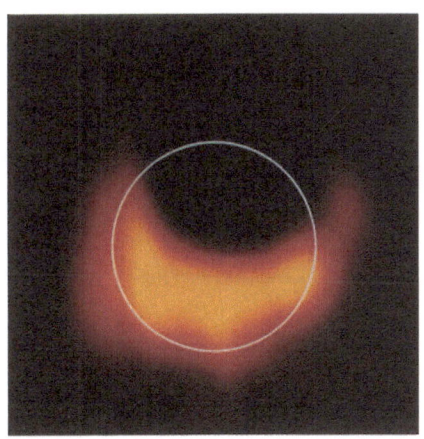

14 15 UT

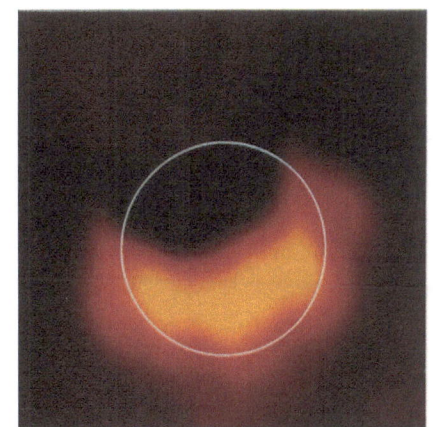

14 40 UT

Preface

It is ever more evident that the major advances of recent years in the understanding of the physics of the sun and the heliosphere are based upon observations made simultaneously, using different techniques. For numerous studies, radio observations play the major role. They are able to probe different layers of the solar atmosphere from the upper chromosphere all the way to the interplanetary medium, and they give precious diagnostics about the physical conditions that dominate. It is essential that these diagnostics are integrated into any global view of the phenomena. Their comprehension requires advanced theoretical efforts and intensive modelling.

The committee CESRA, which represents the community of European solar radio astronomers, decided to organize a series of three successive workshops concentrated on *coronal physics based on radio and space observations*. The first of these workshops took place at Nouan le Fuzelier, France, from 3 to 7 June 1996. The scientific organizing committee was composed of A.O. Benz (Switzerland), A. Krüger (Germany), A. Magun (Switzerland), M. Pick (France), G. Trottet (France), L. Vlahos (Greece), V. Zaitsev (Russia), P. Zlobec (Italy).

In order to stimulate collaborations, the participants comprised all scientific domains whose research overlaps that of radio astronomy. The first workshop had two central objectives: to bring up to date a certain number of problems, and to elucidate the themes to be treated during following workshops. In order to attain these objectives, plenary sessions were organized with all attendees participating. In addition, four working groups were convened, each with two leaders. In total there were sixty-eight participants, coming from ten European countries plus China, the United States of America, India, Israel, and Japan.

This volume contains the majority of the invited reviews, a synthesis of the conclusions, and the results presented by each working group. Some instrumental developments for solar radio astronomy are emphasized: those that are being or have recently been developed, and future radio projects of large scale that have far-reaching consequences for the discipline.

The release of energy in the solar corona remains a central topic, one that manifests itself in different forms, with highly different spatial and temporal scales: there are multitudes of phenomena such as bright points, flares, and coronal mass ejections (CMEs). The key, underlying questions are particularly concerned with the heating of the corona and the process of energy release in flares. Much theoretical work and many present observations bear on the fragmentation of that energy storage and release.

Accelerated particles, detected either by their electromagnetic signatures or by in-situ measurements in interplanetary space, have varied origins that are far from being understood. For example, in the corona and the interplanetary medium, the role of CMEs and shock waves in the acceleration process is not clear. The characteristics of particle beams that propagate from the corona to several astronomical units pose many questions relating to their propagation and their stability. The relationship of accelerated particles to their resultant effects in the solar atmosphere needs also to be understood, e.g., the evaporation of chromospheric material and the related radio emissions.

Decisive progress on all of these questions is to be anticipated from combining observations from the ground and from space (ULYSSES, WIND, SOHO,YOHKOH, GRO, GRANAT...).

The present meeting was organized by the Laboratory of "Physique du Soleil et de l'Héliosphère" (CNRS et Observatoire de Paris), with the help of Orléans University and of the National Center for Scientific Research. This workshop, which was a Euroconference, was sponsored by the European Physical Society and by the "Ministère Francais de l'Education Nationale, de l'Enseignement Supérieur, de la Recherche et de l'Insertion Professionnelle".

This meeting received financial support from the "Conseil Régional de la Région Centre", from the Commission of European Communities (DGXII), from the European Space Agency, from the "Ministère des Affaires Etrangères", from the "Ministère de l'Education Nationale, de l'Enseignement Supérieur, de la Recherche et de l'Insertion Professionnelle", from the National Center for Scientific Research and from the Paris Observatory. I thank all the regional, French and European organisations and a number of colleagues for their help in making the workshop a success. Moreover I gratefully acknowledge A. Raoult President of the Organizing Committee, A. O. Benz President of CESRA, G. Trottet, Editor and a number of referees for evaluating the contributions

Meudon, January 1997   Monique Pick

President of the Scientific Committee

Préface

Il est de plus en plus évident que les étapes importantes franchies au cours des récentes années dans la compréhension de la physique solaire et de l'héliosphère ont reposé sur des observations simultanées faisant appel à différentes techniques d'observations. A cet égard les observations radios jouent un rôle primordial dans de nombreux programmes. En effet, elles permettent de sonder les différentes couches de l'atmosphère solaire, depuis la chromosphère supérieure jusqu'au milieu interplanétaire, et apportent ainsi des diagnostics précieux sur les conditions physiques qui y règnent. Ces diagnostics doivent être intégrés dans une vision globale des phénomènes et leur compréhension nécessite également des efforts théoriques et de modélisation importants.

Le bureau du CESRA qui représente la communauté des radioastronomes solaires européens a decidé d'organiser une série de trois ateliers successifs consacrés à la *physique coronale à partir des observations radios et spatiales*. Le premier atelier s'est tenu à Nouan le Fuzelier, France, du 3 au 7 juin 1996. Le comité d'organisation scientifique comprenait A.O. Benz (Suisse), A. Krüger (Allemagne), A. Magun (Suisse), M. Pick (France), G. Trottet (France), L. Vlahos (Grèce), V. Zaitsev (Russie), P. Zlobec (Italie).

Afin de stimuler les collaborations, la communauté rassemblée comprenait également des scientifiques dont les domaines de recherche interagissent avec la radioastronomie. Ce premier atelier a eu deux objectifs essentiels: faire le point sur un certain nombre de problèmes et dégager les thèmes qui seront traités au cours des deux prochains ateliers. Pour atteindre ces deux objectifs, des séances plénières avec essentiellement des conférences invitées ont été organisées. Les participants ont d'autre part, constitué quatre groupes de travail avec chacun deux animateurs. Il y avait soixante huit participants en provenance de dix pays européens mais également de Chine, des Etats Unis, d'Inde, d'Israel et du Japon.

Ce volume contient la plupart des revues invitées ainsi que la synthèse des réflexions et résultats présentés dans chaque groupe de travail. Il nous a semblé de plus important de faire figurer certains développements instrumentaux en radioastronomie solaire, en cours ou achevés récemment, ainsi que les projets radios de grande envergure pour la discipline.

La libération d'énergie dans la couronne solaire se manifeste sous différentes formes avec des échelles spatiales et temporelles variées: multitude de microphénomènes tels que les points brillants, les éruptions, les éjections de matière coronale (CME). Les questions clés sous-jacentes concernent en particulier le problème du chauffage de la couronne et les processus de libération d'énergie. De nombreux travaux théoriques et expérimentaux portent actuellement sur la fragmentation du stockage et de la libération de cette énergie.

Les particules accélérées qui sont détectées soit par leurs signatures électromagnétiques, soit par des mesures in-situ dans le milieu interplanétaire ont des origines variées qui sont loin d'être comprises. Ainsi, dans la couronne et le milieu interplanétaire, le rôle des CMEs et des ondes de choc dans les processus d'accéleration n'est pas clair. La mise en évidence de faisceaux de particules qui se propagent depuis la couronne jusqu'à des distances de plusieurs unités astronomiques pose de nombreuses questions telles que la compréhension de leur propagation, de leur stabilité. Le lien des particules accélérées avec les effets qu'elles engendrent dans l'atmosphère solaire, tels que l'évaporation chromosphérique ou les émissions radios, doit également être compris.

Sur toutes ces questions, des progrès décisifs sont attendus grâce à la combinaison des observations spatiales (ULYSSES, WIND, SOHO, YOHKOH, GRO, GRANAT...) et des observations obtenues sur Terre.

Cet atelier qui était organisé par le laboratoire de Physique du Soleil et de l'Héliosphère (CNRS et Observatoire de Paris), avec l'aide de l'Université d'Orléans et du Centre National de la Recherche Scientifique était une Euroconférence. Elle était parrainée par la Sociéte Européenne de Physique et par le Ministère Français de l'Education Nationale, de l'Enseignement Supérieur, de la Recherche et de l'Insertion professionnelle.

Cette conférence a été organisée grâce au soutien financier du Conseil Régional de la Région Centre, de la Commission de la Communauté Européenne (DGXII), de l'Agence Spatiale Européenne, du Ministère des Affaires Etrangères, du Ministère de l'Education Nationale, de l'Enseignement Supérieur , de la Recherche et de l'Insertion Professionnelle, du Centre National de la Recherche Scientifique et de l'Observatoire de Paris. J'adresse tous mes remerciements à l'ensemble des organismes régionaux , français et européens ainsi qu'à mes nombreux collègues qui ont apporté leur concours pour l'organisation de cet atelier. Que soient remerciés tout particulièrement A. Raoult Président du comité local d'organisation, A.O. Benz Président du CESRA, G. Trottet responsable de l'édition de ce volume et les rapporteurs qui ont évalué les papiers publiés.

Meudon, Janvier 1997   Monique Pick

Président du Comité scientifique

Contents

Part IV   Radio Instrumentation

List of Participants

Alissandrakis, C. E.   calissan@cc.uoi.gr
  University of Ioannina, Department of Physics, Section of Astro-Geophysics, GR-45110 Ioannina, Greece

Arzner, K.   arzner@sun.iap.unibe.ch
  Universitdt Bern, Institut für Angewandte Physik, Sidlerstrasse 5, CH-
  3012 Bern, Switzerland

Aschwanden, M.J.   markus@astro.umd.edu
  University of Maryland, Astronomy Department, College Park MD 20742,
  USA

Aurass, H.   haurass@aip.de
  Astrophysikalisches Institut Potsdam, Observatorium für solare Radioastronomie, Telegrafenberg A31, D-14473 Potsdam, Germany

Bastian, T.S.   tbastian@nrao.edu
  NRAO, P.O. Box 0, Socorro NM 87801, USA

Bentley, R.   rdb@mssl.ucl.ac.uk
  Mullard Space Science Laboratory, Holmbury St Mary, Dorking, Surrey
  RH5 6NT, UK

Benz, A.   benz@astro.phys.ethz.ch
  Institute of Astronomy, Radio Astronomy Group, ETH-Zentrum, CH-
  8092 Zürich, Switzerland

Biesecker, D.   dab@star.sr.bham.ac.uk
  University of Birmingham, School of Physics and Space Research, Edgbaston, Birmingham B15 2TT, UK

Bogod, V.   vbog@radiosun.spb.su
  Special Astrophysical Observatory of Russian Acad. Sci., St Petersburg
  branch of (SAO-RAS), 196140 St Petersburg, Russia

Braun, R.   braun@nfra.nl
  NRFA, Postbus 2, NL-7990 AA Dwingeloo, The Netherlands

Buttighoffer, A.   Anne.Buttighoffer@obspm.fr
  Observatoire de Paris, DASOP, 5 Place Jules Janssen, F-92195 Meudon
  Cedex, France

Chertok, I. M.   ichertok@lars.izmiran.troitsk.su
  Institute of Terrestrial Magnetism Ionosphere & Radio Wave Propagation, Izmiran, 142092 Troitsk, Moscow Region, Russia

Chiuderi Drago, F.   fdrago@arcetri.astro.it
Universita di Firenze, Dipartimento di Astronomia e Scienza Dello Spazio,
Largo Enrico Fermi 5, I-50125 Firenze, Italy

Coulais, A.   coulais@obspm.fr
Observatoire de Paris, DASOP, 5 Place Jules Janssen, F-92195 Meudon
Cedex, France

Crosby, N.   crosby@obspm.fr
Observatoire de Paris, DASOP, 5 Place Jules Janssen, F-92195 Meudon
Cedex, France

Csillaghy, A.   csillag.@astro.phys.ethz.ch
Institute of Astronomy, Radio Astronomy Group, ETH-Zentrum, CH-
8092 Zürich, Switzerland

Delouis, J.M.   delouis@mesioa.obspm.fr
Observatoire de Paris, DASOP, 5 Place Jules Janssen, F-92195 Meudon
Cedex, France

Dulk, G.A.   dulk@obspm.fr
Observatoire de Paris, DESPA, 5 Place Jules Janssen, F-92195 Meudon
Cedex, France

Einaudi, G.   einaudi@astr14pi.difi.unipi.it
University of Pisa, Physics Department, Piazza Torricelli 2, I-56126 Pisa,
Italy

Fu Qi-Jun   fuqj@bao01.bao.ac.cn
Beijing Astronomical Observatory, Chinese Academy of Science, 100080
Beijing, China

Gelfreikh, G.B.   gbg@saoran.spb.su
Main (Pulkovo) Astronomical Observatory, Russian Academy of Sciences,
196140 St Petersburg, Pulkovo, Russia

Gopalswamy, N.   gopals@astro.umd.edu
University of Maryland, Department of Astronomy, College Park MD
20742, USA

Grebinskij, A.S.   agreb@vag.stud.pu.ru
St Petersburg University, Institute of Radio Physics, Ulyanovskaya Street
1, 198904 St Petersburg/Petrodvorets, Russia

Hackenberg, P.   phackenberg@aip.de
Astrophysikalisches Institut Potsdam, Observatorium für solare Radioastronomie, Telegrafenberg A31, D-14473 Potsdam, Germany

Haerendel, G.   hae@mpe-garching.mpg.de
Max-Planck-Institut für Extraterrestrische Physik Giessenbachstraße D-
85740 Garching, Germany

Hénoux, J-C.   henoux@obspm.fr
Observatoire de Paris, DASOP, 5 Place Jules Janssen, F-92195 Meudon
Cedex, France

Hildebrandt, J.   jhildebrandt@aip.de
Astrophysikalisches Institut Potsdam, An der Sternwarte 16, D-14482
Potsdam, Germany

Huang, G.   pmoyl@bao01.ac.cn
Purple Mountain Observatory, Nanjing 210008, China

Jiricka, K.   jiricka@asu.csa.cz
Astronomical Institute, CS-25165 Ondrejov, Czech Republic

Karlicky, M.   karliky@asu.cas.cz
Astronomical Institute, CS-25165 Ondrejov, Czech Republic

Kerdraon, A.   kerdraon@obspm.fr
Observatoire de Paris, DASOP, 5 Place Jules Janssen, F-92195 Meudon
Cedex, France

Klassen, A.   aklassen@aip.de
Astrophysikalisches Institut Potsdam, Observatorium für solare Radioastronomie, Telegrafenberg A31, D 14473 Potsdam, Germany

Klein, K.-L.   klein@obspm.fr
Observatoire de Paris, DASOP, 5 Place Jules Janssen, F-92195 Meudon
Cedex, France

Kliem, B.   bkliem@aip.de
Astrophysikalisches Institut Potsdam, An der Sternwarte 16, D-14482
Potsdam, Germany

Krasnosselskikh, V.   vkrasnos@cnrs-orleans.fr
L.P.C.E., 3A, Avenue de la Recherche Scientifique, F-45071 Orlians Cedex
2, France

Krucker, S.   krucker@astro.phys.ethz.ch
Institute of Astronomy, Radio Astronomy Group, ETH-Zentrum, CH-
8092 Zürich, Switzerland

Kundu, M.R.   Kundu@astro.umd.edu
University of Maryland, Department of Astronomy, College Park MD
20742, USA

Leblanc, Y.   leblanc@obspm.fr
Observatoire de Paris, DESPA, 5 Place Jules Janssen, F-92195 Meudon
Cedex, France

Lefeuvre, F.   lefeuvre@cnrs-orleans.fr
L.P.C.E., 3A, Avenue de la Recherche Scientifique, F-45071 Orlians Cedex
2, France

Lin, R.P.   boblin@sunspot.ssl.berkeley.edu
University of California, Space Sciences Laboratory, Berkeley CA 94720,
USA

Magun, A.   magun@sun.iap.unibe.ch
Universitdt Bern, Institut für Angewandte Physik, Sidlerstrasse 5, CH-
3012 Bern, Switzerland

Maia, D.   dalmiro@mesioq.obspm.fr
Observatoire de Paris, DASOP, 5 Place Jules Janssen, F-92195 Meudon Cedex, France

Mann, G.   gmann@aip.de
Astrophysikalisches Institut Potsdam, Observatorium für solare Radioastronomie, Telegrafenberg A31, D-14473 Potsdam, Germany

Manoharan, P.K.   manoharan@obspm.fr
Observatoire de Paris, DASOP, 5 Place Jules Janssen, F-92195 Meudon Cedex, France

Mein, P.   meinp@obspm.fr
Observatoire de Paris, DASOP, 5 Place Jules Janssen, F-92195 Meudon Cedex, France

Melnikov, V.F.   meln@nirfi.sci-nnov.ru
Radiophysical Research Institute, B. Pecherskaya St. 25, 603600 Nizhnii Novgorod, Russia

Mercier, C.   mercier@obspm.fr
Observatoire de Paris, DASOP, 5 Place Jules Janssen, F-92195 Meudon Cedex, France

Nakajima, H.   nakajima@nro.nao.ac.jp
Nobeyama Radio Observatory, Minamimaki, Minamisaku, Nagano 384-13, Japan

Pick, M.   pick@obspm.fr
Observatoire de Paris, DASOP, 5 Place Jules Janssen, F-92195 Meudon Cedex, France

Pohjolainen, S.   spo@clan.hut.fi
Helsinki University of Technology, Metsdhovi Radio Research Station, Metsdhovintie 114, FIN-02540 Kylmala, Finland

Poquérusse, M.   poquerusse@obspm.fr
Observatoire de Paris, DESPA, 5 Place Jules Janssen, F-92195 Meudon Cedex, France

Pustilnik, L.   lev@yarden.yarden.ac.il
Academic College, Solar Radio Observatory of Jordan Valley, Israkl

Raoult, A.   raoult@obspm.fr
Observatoire de Paris, DASOP, 5 Place Jules Janssen, F-92195 Meudon Cedex, France

Raulin, J.P.   raulin@astro.umd.edu
University of Maryland, Department of Astronomy, College Park MD 20742, USA

Schmieder, B.   schmieder@obspm.fr
Observatoire de Paris, DASOP, 5 Place Jules Janssen, F-92195 Meudon Cedex, France

Schwarz, U.   schwarz@agnld.Uni-Potsdam.de
Universitdt Potsdam, Institut für Theoretische Physik & Astrophysik, PF 601553, D-14415 Potsdam, Germany

Shibasaki, K.   shibasaki@nro.nao.ac.jp
Nobeyama Radio Observatory, Minamimaki, Minamisaku, Nagano 384-13, Japan

Simnett, G.   gms@star.sr.bham.ac.uk
University of Birmingham, School of Physics and Space Research, Edgbaston, Birmingham B15 2TT, UK

Stehling, W.   stehling@astro.phys.ethz.ch
Institute of Astronomy, Radio Astronomy Group, ETH-Zentrum, CH-8092 Zürich, Switzerland

Treumann, R.A.   tre@mpe-garching.mpg.de
Max-Planck-Institut für Extraterrestrische Physik, Karl-Schwartzschild Str.1, D-85748 Garching, Germany

Trottet, G.   trottet@obspm.fr
Observatoire de Paris, DASOP, 5 Place Jules Janssen, F-92195 Meudon Cedex, France

Vilmer, N.   vilmer@obspm.fr
Observatoire de Paris, DASOP, 5 Place Jules Janssen, F-92195 Meudon Cedex, France

Vlahos, L.   vlahos@helios.astro.auth.gr
University of Thessaloniki, Physics Department, GR-54006 Thessaloniki, Greece

Walsh, R.W.   robert@dcs.st-and.ac.uk
University of St. Andrews, School of Mathematical and Computational Sciences, KY16 9SS Scotland, UK

Zaitsev, V.   za130@appl.sci-nnov.ru
Institute of Applied Physics, Russian Academy of Sciences, Ulyanova 46 St, 603600 Nizhny Novgorod, Russia

Zlobec, P.   zlobec@ts.astro.it
Trieste Astronomical Observatory, Via G.B. Tiepolo 11, I-34131 Trieste, Italy

Zlotnik, E.   zlot@appl.sci-nnov.ru
Institute of Applied Physics, Russian Academy of Sciences, Ulyanova 46 St, 603600 Nizhny Novgorod, Russia

Zmoos, C.   zmoos@astro.phys.ethz.ch
Institute of Astronomy, Radio Astronomy Group, ETH-Zentrum, CH-8092 Zürich, Switzerland

Part I

Energy Release in the Quiet and Active Solar Atmosphere

Diagnostics of Energy Release
in the X-Ray Corona

Robert D. Bentley

Mullard Space Science Laboratory, University College London,
Holmbury St. Mary, Dorking, Surrey RH5 6NT, United Kingdom

Abstract. Energy is released in the corona in many different forms. In this paper we discuss phenomena that are observed in X-rays in flares and sub-flares. Recent results of X-ray bright-points, transient brightenings and jets are reviewed, and diagnostics that result from bulk plasma motions are discussed. Finally, three events are examined in detail and the implications of the observation discussed in relation to flare models.

Résumé. Il y a beaucoup de façons de libérer de l'énergie dans la couronne solaire. Dans cet article, nous examinons les phénomènes qui sont observés en rayons X pendant les éruptions et les sous-éruptions. Nous passons en revue les résultats récents obtenus sur les points brillants X, les embrillancements transitoires et les "jets". Nous discutons également les diagnostics qui résulent des mouvements macroscopiques du plasma. Finalement, nous présentons une analyse détaillée de trois événements. Les résultats de cette analyse sont discutés dans le cadre de différents modèles d'éruption.

1 Introduction

The release of energy in the solar corona produces many spectacular structures that can be observed in soft X-rays. Huge loops can be seen connecting distant points on the solar surface and extending far into the corona. These represent tubes of hot plasma confined by the coronal magnetic field and sustained by releases of energy from the lower layers of the solar atmosphere.

Although there are clearly large amounts of energy released when large-scale structures change their form or erupt, we will confine this paper to smaller, more concentrated forms of energy release since the current generation of X-ray instrumentation are better suited to provide detailed information on such phenomena. Also, because of limitations in the space available, this discussion will not cover the changes in the energy release as activity varies over the solar cycle.

The discussion therefore centres mainly on phenomena associated with flares and sub-flares. At the lower end of the scale, we discuss recent observations of brightpoints, transient brightenings and jets. Such observations mainly involve images gathered by the *Yohkoh* Soft X-ray Telescope (SXT; Tsuneta et al. 1991) since they involve such small releases of energy that they fall below the sensitivity threshold of other instruments. As we look at larger events, instruments such as the *Yohkoh* Bragg Crystal Spectrometer (BCS; Culhane et al. 1991), and other similar instruments on previous missions, are used to determine the detailed information about the flaring plasma. Here we examine the bulk motion of plasma determined from the width and Doppler-shift of spectral lines, particularly during the onset of flares. For such larger events, it is also possible to determine information about the production of hard X-ray - these are normally considered to originate from sites where energy is being deposited. For this we include results from the *Yohkoh* Hard X-ray Spectrometer (HXT; Kosugi at al. 1991).

Finally, we will review three events that have been studied in detail and examine what the observations mean for the models of energy release.

2   Features Associated with Sub-Flare Activity

It is open to debate where some of the phenomena discussed here fit in relation to flares. At one time flares were only thought of as quite large releases of energy, but as the sensitivity of instruments has improved, we have seen ever smaller features that are "flare-like" in structure if not stature. The frequency of such features is often greater than the larger "flares" and the total energy that they release is considerable, although the area, and rate per unit area of the energy dissipation are very different for those of a flare. It is now thought that such features may just represent the low end of a whole range of phenomena related to energy release. In this context, we discuss X-ray bright-points, transient brightenings, and jets.

2.1   X-ray Bright-Points

X-ray Bright-Points (XBPs) were first studied in detail using AS&E *Skylab* X-ray images. They are characterized as compact structures, typically less than 60 arcseconds across, with lifetimes from minutes to a few hours (Golub et al., 1976, and many other references). Other than the *Skylab* data, studies have mainly involved data taken during rocket flight lasting a few minutes. *Skylab* itself only provided good coverage for limited time periods and it was only following the successful launch of *Yohkoh* that a continuous, long-term X-ray observations became available.

During the *Yohkoh* mission there have been several coordinated observing campaigns to explore the occurrence and variability of bright-points on the timescales of hours. Ground based observations were used to provide information on the underlying magnetic fields and chromospheric structures (e.g. Fig. 1). The studies have been summarized by Harvey (1996) who reported that the disappearance of surface magnetic flux is an important process related to the occurrence of bright-points, although a number of bright, small and short-lived bright-points overlie no apparent magnetic structure or unipolar magnetic network. Harvey found that more than half of the bright-points show gradual (one to ten minutes) variability, and that just over a third of XBPs flared, exhibiting intensity changes on timescales of seconds to minutes and often showing multiple pulses during an event. All bright-points were found to have associated He I 1083 dark points (although the inverse is not true) - the dark points are thought to be a chromospheric maker of cancelling magnetic dipoles (Harvey 1985).

Harvey concluded that the emergence or cancellation of photospheric magnetic flux is not in itself a necessary and sufficient condition for the occurrence of X-ray bright-points, and that the occurrence and variability of the bright points result from the interaction and (often sporadic) reconnection of the emerging and cancelling magnetic fields with existing, local magnetic fields.

Kundu et al. (1994) have found an association between weak metric type III radio bursts and flaring XBPs. Since type III bursts are produced by non-thermal beams of electrons, this association represents strong evidence that the XBP-flare mechanism is capable of accelerating particles to non-thermal energies, as well as producing the heated material detected in soft X-rays.

2.2 Transient Brightenings

The *Yohkoh* SXT shows numerous transient brightenings of compact coronal loops (Shimizu et al. 1992; Shimizu et al. 1994; Shimizu 1996). The energy involved in such brightenings is in the range 10^{25} to 10^{29} ergs. The lower end of the range is just above the energy associated with nano-flares and the upper end similar to that associated with small flares. While the more intense brightenings can be identified as small soft X-ray enhancements (below B-class) in *GOES* time profiles, the weaker ones are not detectable and represent the smallest transient energy releases in X-rays detected so far.

Transient brightenings show great variations in morphology; they are mostly in the form of single or multiple loops, although some point-like brightenings are also seen. The loops tend to brighten from their footpoints (or the site where loops are apparently in contact) with the entire loop eventually brightening. Shimizu et al. (1994) concluded that the brightenings are probably caused by sudden releases of energy due to magnetic reconnection.

Observations by *Yohkoh* of transient brightenings allowed the frequency distribution of subflares to be determined for the first time. Shimizu (1995) found that as a function of peak flux, the brightenings can be represented by a power law with an index of 1.4 ± 0.1 over 2 orders of magnitude (Fig. 2), and as a function of energy release, by a power law of 1.5–1.6 in the energy range greater than 10^{27} ergs. The power law of the latter relationship is flatter that the slope of hard X-ray peak flux of flares (1.7–1.8; e.g. Dennis 1985), but is in good agreement with that of the total flare energy calculated by integrating the thick-target energy rate over the flare duration (1.53; Crosby, Aschwanden and Dennis 1993).

Shimizu (1996) concludes that transient brightenings are the low-energy extension of the general flare distribution and that the superposition of the

6   Robert D. Bentley

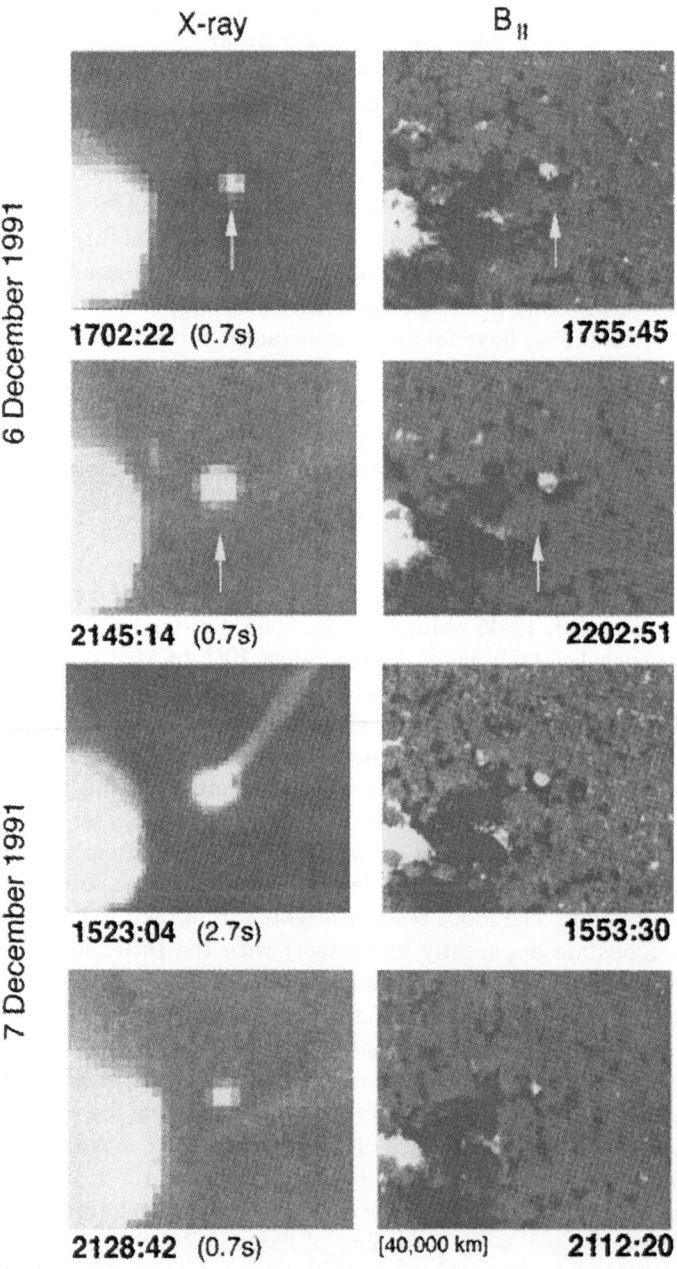

Fig. 1. Comparison of *Yohkoh* SXT images (left) with magnetograms (right) of an XBP associated with a small emerging active region, indicated by an arrow in the top panel. As the region expanded, its positive (white) pole cancelled with adjacent negative polarity (black) magnetic network (arrow in the second panel). The XBP flares at 15:07 UT on 1991, December 7 and was associated with an X-ray jet (shown in the third panel) (after Harvey 1996).

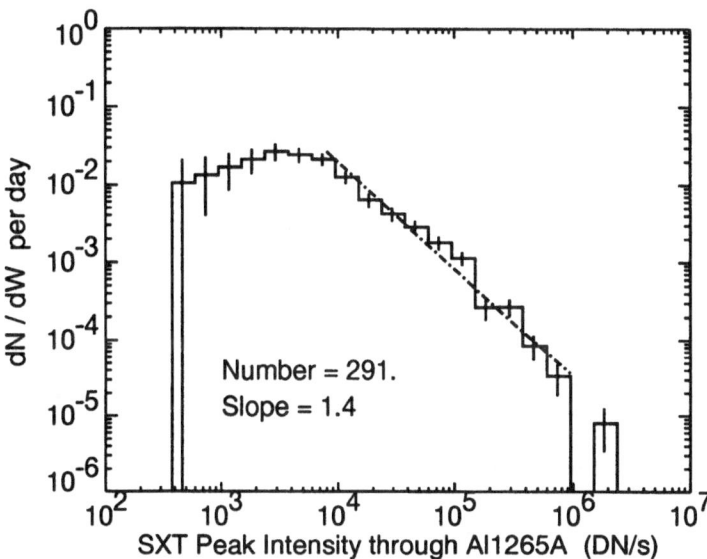

Fig. 2. Frequency distribution of transient brightenings as a function of soft X-ray peak intensity (after Shimizu 1996).

transient brightenings does not contribute significantly to the energy required to heat the corona. Shimizu speculates that, at the lower energy end, the transient brightenings seen by *Yohkoh* are really just the bigger of the nano-flares and that the corona may be heated by a very large number of (yet to be detected) smaller nano-flares.

2.3 Jets

Soft X-ray jets are a discovery of *Yohkoh*; they are transitory X-ray enhance-ments with apparent collimated motion (Shibata et al, 1992). Jets were prob-ably not noticed before because they are so short-lived and can only be seen in the high cadence observations of the type produced by the *Yohkoh* SXT. They are associated with microflares or subflares which occur in X-ray bright-points, emerging flux regions, or active regions.

A statistical study of 100 jets (Shimojo et al., 1996) showed that typically the length of the jets is 10^3–4×10^5 km (with an average of $\simeq 1.5 \times 10^5$ km), the apparent velocity is 10–1000 km s^{-1} (with an average of $\simeq 200$ km s^{-1}), and the lifetime ranges from a few minutes to more than a few hours. The number of jets decreases as the length, velocity or lifetime increases (a similar pattern is seen with flares). The temperature of jets is comparable to that of microflares at the footpoints of the jet ($\sim$ 3–6 MK), the electron density is 3×10^8–3×10^9 cm^{-3}, and the kinetic energy is estimated to be 10^{25}–10^{28} erg. An example of a jet can be seen in the third panel of Fig 1.

Shibata et al. (1996) discuss many aspects of the observations and theory of jets, and speculate that X-ray jets, and the X-ray plasma ejections (or plasmoids) seen in some large flares, result from the same phenomena. The only difference is that while plasmoids are produced (and ejected) by the reconnection process, X-ray jets are caused by the enhanced gas pressure that is found behind the fast shock produced by the reconnection. Kundu et al. (1995) have found type III radio bursts (over a range of frequencies) that are spatially and temporally coincident with X-ray jets. This association implies the process that produces the jet involves both the acceleration of electrons to several keV and the heating of plasma responsible for the soft X-rays. It also implies that there are open field lines in dense coronal structures.

Shibata et al. suggest that X-ray jets may provide clues to understanding coronal heating and the acceleration of the high speed solar wind. They base this on the study of Feldman et al. (1993) who found that the velocity distribution in the solar wind was not consistent with the ambient flow from coronal holes, but could be explained by the plasma being both heated and accelerated by the the episodic generation of jets driven by the reconnection of newly emerged bipolar magnetic field with elements of the chromospheric network.

3   Observational Evidence of Mass Motions in Flares

The phenomena described so far have been observed in image data and were seen a consequence of having an advanced X-ray telescope, the *Yohkoh* SXT, in orbit for an extended period. There are other signatures of energy release in the corona that have been observed for many years, and from many different spacecraft. These are primarily associated with flares and here we discuss aspect of the emission that represent the signatures of mass motions.

X-ray spectrometers have been flown on several missions over the last 20 years: *SMM*, *P78-1*, *Hinotori*, and *Yohkoh*. Such spectrometers use crystals to diffract incident X-ray photons and sort them by wavelength (in the ultra-violet a grating performs the same function) allowing high resolution measurements of spectral emission lines to be made. Although technology may evolve in the future, such a technique has been necessary up to now because detectors that work at these energies do not have the energy resolution that would allow them to sort the photons directly.

At X-ray wavelengths, the emission lines observed are produced by highly ionized plasma of ~ 1 MK and upwards. Almost all of the electrons have been stripped from the atom and the electrons orbiting the remaining nucleus make transitions that are similar to the hydrogen and helium atoms - as such they are described as hydrogen- and helium-like ions. In the *Yohkoh* BCS, the lines of Fe XXV, Ca XIX and S XV are helium-like, and those of Fe XXVI are hydrogen-like. The spectra observed by the BCS are analysed using spectral fitting programmes; these determine the intensity and width of the lines and can also fit multiple components if required. The derived line intensities are used to determine the emission measure and electron temperature; the latter is derived from the ratio of the resonance and dielectronic satellite lines (Bely-Dubau at al, 1982). Asymmetries in the lines can be used to determine the evolution of the velocity and emission measure of the blueshifted component.

When the *Yohkoh* BCS was being designed, it was decided to improve its sensitivity over previous spectrometers in order to be able to observe Doppler shifts and line broadening earlier during the onset of the flare. Largely as a consequence of not using a collimator, the *Yohkoh* BCS was 5 – 10 times more sensitive than the *SMM* BCS; this has allowed the instrument to observe much smaller flares than before and events as small as the B1–B3 (*GOES*-class) have been studied. A large fraction of the soft X-ray events observed by the BCS have counterparts in hard X-rays. The smaller events often do not, but this may purely be a matter of the sensitivity of the hard X-ray instruments. This must be borne in mind when studying the relationship of the hard and soft X-ray signatures.

3.1   Blueshifts

At the onset of some flares, an excess emission is seen on the blue (short-wavelength) side of soft X-ray emission lines such as those of the helium-like ions of Ca XIX and Fe XXV (e.g. Fig 3). Blueshifts of this type were first observed by instruments on *SMM* and *P78-1*, and are thought to result from Doppler-shifted material moving towards the observer in what appears to be directed motion. They are considered to be the signature of *Chromospheric Evaporation*, the upflow of material from the chromosphere into the corona resulting from massive localized heating.

The blue component is seen at a time that is closely related to emission in hard X-rays. In an attempt to better characterize the component, and thus to understand the source of the heating that drives the evaporation, we will discuss the velocity, timing relative to hard X-rays, locations and temperature dependence of the component.

Velocity of the Blueshifted Component The velocity of the blueshifted component varies considerably from flare-to-flare, ranging from a few tens to several hundred kilometres per second. The maximum velocity observed does not appear to be sensitive to the magnitude of the peak hard X-ray flux (i.e. the energy flux of the precipitating energetic electrons), but may be dependent on the structure of the hard X-ray burst.

At the onset of most soft X-ray flares, a stationary (or "rest") component of the spectrum is observed; the blueshifted material causes an asymmetry or wing to this component. In a few flares only the blueshifted component is seen and in such cases the velocity of this component is seen to decrease in magnitude towards the "rest" position during impulsive phase. Very rarely,

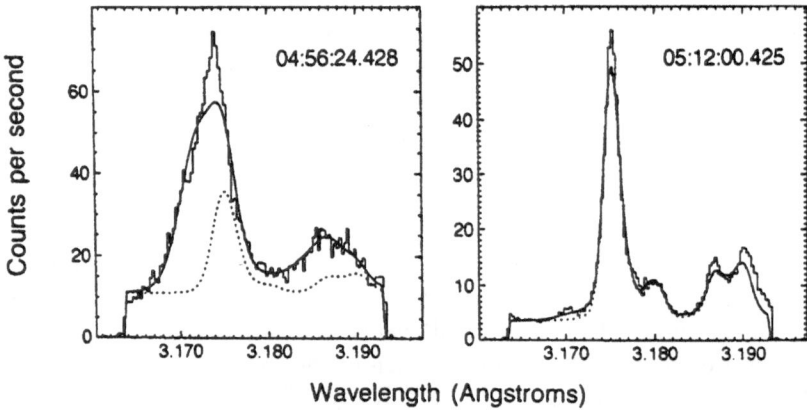

Fig. 3. Ca XIX spectra recorded by the *Yohkoh* BCS during the flare of 1991, December 16. A spectrum taken early in the flare (left) is compared with one during the decay phase (right); the results of the spectral fit are shown. The line centroid is clearly blueshifted in the earlier spectrum; the shift corresponds to a velocity of 280 km s^{-1} (after Culhane at al., 1994).

the blue component is completely separate from the stationary component and is thought to represent an ejected plasmoid (Bentley et al., 1986).

That the bluewing is emitted by material flowing up the legs of flaring loops within the corona is confirmed by the centre-to-limb (*cosine*) variation in the velocity of the blueshifted component (e.g. Mariska et al. 1993). Such a dependency arises because the loop legs should normally be radial to the solar surface, and loops that are located near the solar limb would not be expected to present a velocity component to the observer.

It should be noted that velocities quoted for the blue component in many published studies are determined by model dependent codes - a technique that fits two Gaussian components is most frequently used (e.g. Antonucci et al., 1982; Fludra et al., 1989). As the flare progresses there are inherent uncertainties using such techniques in determining what proportion of the signal should be attributed to the moving and stationary components. This causes particular problems when the blue component is weak near the end of the impulsive phase and the codes continue to attribute a width and intensity to the component after it has lost any real significance. Although there have been various attempts to improve the fitting procedures, e.g. by fitting three components (Antonucci et al., 1990), or by coupling the blue and stationary components together in some way (Fludra et al., 1989), the basic difficulty remains that such techniques poorly describe the problem since the blue component clearly consists of more than a simple Gaussian distribution of velocities. This is in itself interesting since it probably indicates that the blueshift does not represent a single movement of material, but rather the integral of a large number of discrete directed motions. Whether this is indicative of flows of different velocities up many individual loops, or an inhomogeneous flow within a single loop is not clear, but a study of simple, compact flares may yield more information on the issue.

Timing of the Blue Component Whether the hard X-rays peak before the soft X-ray bluewing has been a matter of controversy. Investigators have disagreed on the timing of the two (e.g. Bentley et al., 1994; Doschek et al., 1993; Cheng et al., 1994; Plunkett et al., 1994), but views are converging. This relationship is of great significance since it places severe constraints on any flare models.

Bentley et al. (1994) find that the principal blue component peaks almost the same time or shortly after a hard X-ray burst. The burst is not necessarily the largest hard X-ray spike in an event, but it is usually the first significant one. Bentley (1996) found that some differences in the conclusions reported by investigators may have arisen because of the analysis techniques used, and that a tendency to concentrate on the largest hard X-ray peak has caused confusion. The scaling of the intensity of the bluewing to that of the stationary component will produce a curve whose evolution is dependant on the rate of change of the more intense stationary component. Consequently, blueshifts may have been erroneously perceived to peak before the hard X-rays. Weak blue-wing emission is sometimes seen before the onset of the impulsive phase, but this is of very low intensity compared to the main blueshifted component and appears to be associated with low-level hard X-ray activity.

Winglee et al. (1991) studied the relationship between hard and soft X-rays using *SMM* data. For all flares they show that the hard X-rays have a power-law spectrum that breaks down during the rise phase and beginning of the decay phase, after which the spectrum changes to either a single power law or a power-law that steepens at high energies. They also find that the maximum velocity of the blueshifted component occurs in association with the disappearance of the (downward) break in hard X-ray spectrum.

At some point, the blue component ceases although the hard X-ray emission has not always finished. Why this happens is not clear. Some investigators suggest that it is because the loop has filled to capacity and further evaporation is inhibited by density and pressure consideration. However, Winglee et al. (1991), think that the hard X-ray and blueshifted emission are both produced by quasi-static electric fields that disappear when the hot plasma reaches the chromosphere and enhances the cross-field conductivity in that region and thus reducing the electric fields needed to close the current system.

Location of the Upflows It is only possible to determine the location of an upflow by indirect means. The crystal spectrometers flown to date have not been capable of imaging, and observations by the *Yohkoh* BCS can only roughly tie a source to a flare site. However, from other simultaneous

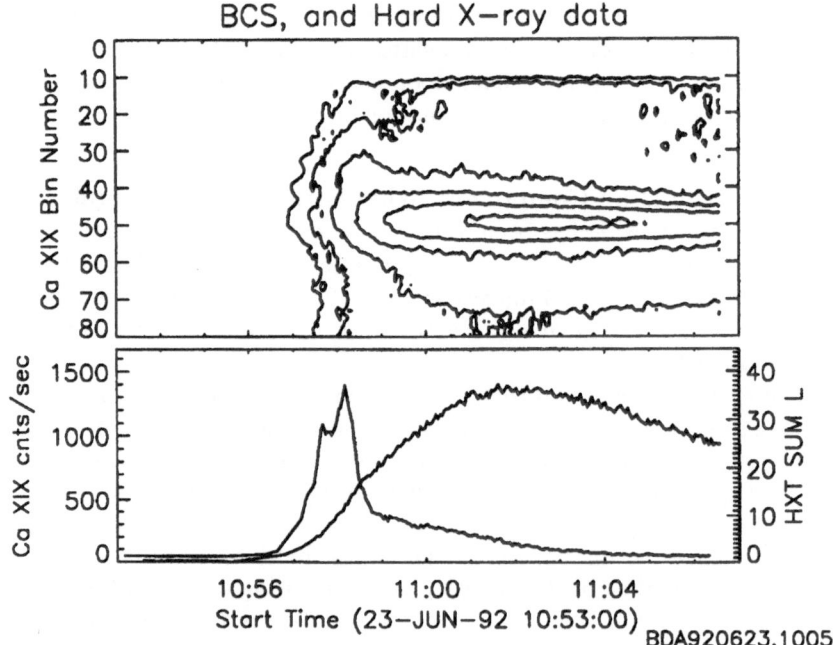

Fig. 4. Relative timing of the SXR bluewing and the HXR burst. The top panel shows Ca XIX spectra from the *Yohkoh* BCS plotted against time; lower bin numbers represent shorter wavelengths. The lower panel shows the lightcurves of total counts from Ca XIX (more gradual curve) and the HXT Lo channel (spike at ~10:58 UT). The blue wing can be seen as an asymmetry in the contours in the top plot between 10:58 and 10:59; the maximum velocity appears to be just after the peak in hard X-rays. The stationary component is represented by the ridge in the contours near spectral bin 40 (after Bentley, 1994).

observations it is possible to draw more definite conclusions about where the upflow is taking place.

A comparison of pixel lightcurves[1] from the SXT with the time evolution of the blue-wing seen in the BCS is a good way to identify the site of the blueshifted component - where the time profiles match, that location probably represents the source of the upflow. By comparing the temperatures and emission measures determined from measurements by the BCS and SXT, it is possible to confirm that the volume of the emitting plasma is the same in both cases. If a comparison is also made with the hard X-ray time profile,

---

[1] When making such comparisons, care must be taken in deciding exactly what area in the SXT image should be included. Depending on how an area is selected from the image, if the source moves or grows so that it extends outside of the selected area, the derived time evolution could be affected. Taking the whole flare will usually give a better match with the HXT and BCS.

additional information may be derived (Strong et al., 1994, Savy, 1994). In some simple flares, the hard X-ray flux time profile appears to match the lightcurve of the Ca XIX blue-wing and the time derivative of the SXT fluxes.

Upflows originate from regions where energy is being released in the chromosphere. Depending on whether the site is the same as that of an hard X-ray or Hα footpoint, information about how the energy is being released can be determined. The presence of hard X-ray emission suggests that the energy is released through the precipitation of energetic particles; its absence indicates that the energy is conveyed to the site by some other mechanism. Which mechanism is preferred places constraints on any models of the event.

Temperature Dependence of the Blue Component If we examine the magnitude of the blueshift in soft X-ray lines formed at different temperatures, we can determine information about the heating mechanisms involved. It is generally observed that lines formed at high temperature show larger blueshifts than those formed at lower temperatures (Antonucci et al., 1990). Zarro (1993) studied S XV from the *SMM-FCS* instrument and a comparison of the velocities observed with those of Ca XIX and Fe XXV is given in Table 1. This velocity-temperature relationship suggests a physical link between the plasma upflow motions implied by these blueshifts and the plasma heating that occurs in flares.

Table 1. Temperature dependence of the blueshifted component

Ion	Peak Temperature	Maximum Velocity
Fe XXV	50 MK	≤ 800 km s^{-1}
Ca XIX	30 MK	≤ 500 km s^{-1}
S XV	15 MK	≤ 200 km s^{-1}

Fludra et al. (1989) report that the mean temperatures of the moving and stationary plasma normally agree although at the onset of the rise phase the temperature of the upflowing plasma may be *higher*. They conclude that this implies that the flare energy deposited lower in the loop, rather than at the top.

3.2   Line Broadening

At the onset of many flares, the line-widths of emission lines observed in soft X-rays are in excess of the value predicted by atomic theory. This line-broadening ($v_t = 100 \sim 200$ km s^{-1}) is thought to be non-thermal in origin and is probably caused by non-directed mass motions, or *turbulence*, in the flare region. Here we examine the timing and location of the source of line broadening.

Timing of Line Broadening Line broadening is greatest near flare on-set and is typically observed in time correlation with hard X-ray emission (Tanaka et al., 1982). The turbulence usually continues as long as the hard X-rays are present and it has normally disappeared by the peak of the flare, although there have been reports of turbulence detected in decay phase (e.g. Fludra et al., 1989).

This coincidence in timing suggests that the non-thermal broadening are a direct consequence of the same process that produces the hard X-rays.

Location of the Source of Line Broadening Whereas the size of the blueshifted component varies across the solar disk, non-thermal line broad-ening does not (Saba at al., 1986; Mariska et al., 1993). This suggests that the broadening is not directed in nature, but rather consists of many randomly oriented components.

The line-broadening is thought to be closely related with energy release (Antonucci et al., 1982), and this is normally indicated by the presence of im-pulsive hard X-rays (Winglee at al., 1991). Flare observations by the *Yohkoh* HXT, show that energy is normally released at the loop footpoints, but in some events (Masuda et al., 1994) a hard X-ray source was also found at the top of the loop. The footpoints are thought to be where precipitating en-ergetic electrons are impacting the chromosphere, but Masuda et al. (1994) suggest that the loop-top site is caused by a shock resulting from magnetic reconnection (see Fig 5).

The location of the site of the line-broadening has also been studied using the solar limb to occult different parts of the flare structure. Khan at al. (1995) examined four flares over an 11 hour period as an active region rotated over the limb and reported that several spectral characteristics, including non-thermal broadening, are virtually indistinguishable no matter how much of the lower parts of the loop is obscured. They conclude that the broadening is either independent of height, or the source is placed near the loop top, and that the energy release driving flares is therefore sited in the corona. Mariska et al. (1995) looked at eight flares from different active regions near or beyond the solar limb. In four of the flares, the footpoints (as observed by HXT) were occulted, while in the other four they were clearly visible. They report that if footpoints are obscured, the peak non-thermal velocities are lower and the observed hard X-ray emission is softer.

There are many different models that try to explain non-thermal broad-ening and observations are clearly important in deciding the validity of each model. For example, the model of Alexander (1990) predicts a variation with occulted height. Mariska et al. (1995) appear to be consistent with this, al-though the variations they report mainly relate to the visibility of the foot-

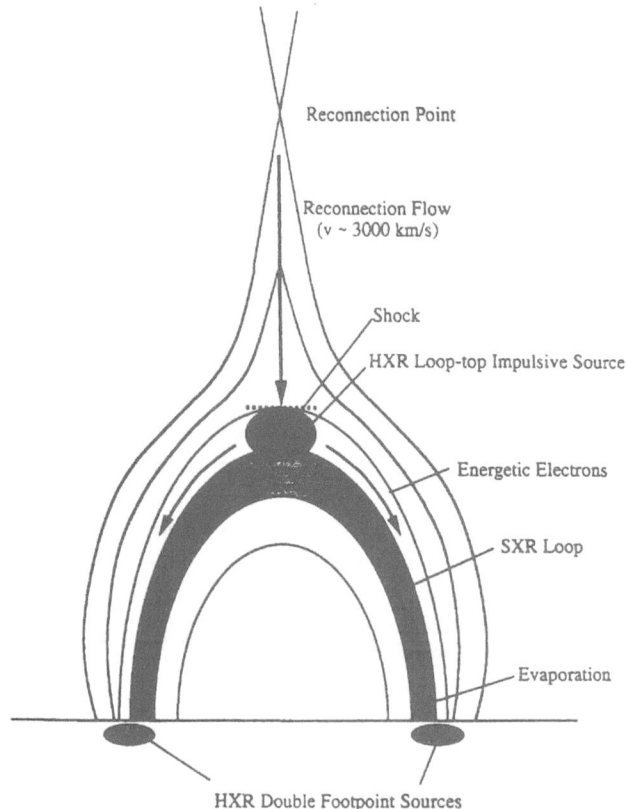

Reconnection Point

Reconnection Flow
(v ~ 3000 km/s)

Shock

HXR Loop-top Impulsive Source

Energetic Electrons

SXR Loop

Evaporation

HXR Double Footpoint Sources

Fig. 5. Schematic showing a model that explains the impulsive the "loop-top" hard X-ray source above the soft X-ray loop. The outflow from the reconnection site forms a shock that created 200 MK plasma when it impinges on the existing closed loop (after Masuda 1993).

point. Khan at al. (1995) do not observe the base of the loop and may miss low-altitude variations, but at higher altitudes their observations do not suggest a variation with height. It is therefore not clear what the variation with height really is and a more detailed analysis is needed.

4   A Detailed Study of Three Events

Yohkoh has observed a large number of flares, and many have been analysed. In Sect.3 we summarized some of the signatures of energy release that are found in data taken by *Yohkoh* and other spacecraft. Here we examine three events in more detail.

All the events include observations of Doppler shifted material by the *Yohkoh* BCS, but conclusions as to what drives the evaporation differ. By examining the energetics of the event, and exploring the signatures seen in hard and soft X-rays we can hope gain an understanding of what drives a given event.

4.1 Event of 1991, November 15

This was the first very large flare observed by *Yohkoh*, and as such has been studied in great detail (Wulser et al., 1994; Canfield et al., 1992, Sakao et al., 1992).

At 22:37 UT on 15 November, 1991, a X1.5/3B flare occurred in NOAA AR 6919, located at S13 W18. Around 10 minutes before the onset of this flare, a filament started moving and it is a smaller event associated with the eruption of the filament that has proved most interesting. During the main flare the *Yohkoh* BCS saturated because the soft X-ray emission was too strong. No significant blueshift was observed during the impulsive phase of this event, even though if present it should have been visible before saturation set in - the reason for this was not determined. At the onset of the smaller event, however, large blueshifts were observed at the same time as redshifts were seen in Hα. The study of Wulser at al. focuses on this period.

Soft X-ray images from SXT were used to measure several preflare and flare properties: the soft X-ray volume, and the preflare temperature and emission measure. Wulser et al. measured the footpoint separation as 1.0×10^9 cm and deduced a loop length of 1.57×10^9 cm; they determined the loop cross-sectional area from soft X-ray images to be 3.8×10^{17} cm^2, giving a total loop volume of 5.9×10^{26} cm^3. The images taken with the Be filter were used since this filter represents the closest approximation available to the Ca XIX spectra. Just before the main event, the bulk of the emission came from a bright elongated feature that connected Hα kernels on either side of the magnetic neutral line; this was interpreted as a coronal loop in projection. A third kernel is seen located at a greater distance from the other two.

The Ca XIX resonance line exhibits a strong blue wing during the period of the earlier flare, and also a small but significant shift in the main component, i.e. the normally stationary component is moving. The velocity of the blue wing component is ~250 km s^{-1}, while that of the main component is ~50 km s^{-1}.

Wulser et al. made a detailed study of the momentum balance of this event. Although the momentum balance had been confirmed for several flares (Canfield et al., 1990), this event represented the first in which it was possible to show that the locations of the upflowing (blueshifted) coronal material and downflowing (redshifted) chromospheric material were the same. Brightpoints were found to be present in the soft X-ray images of SXT, the hard X-ray images of HXT, and the Hα taken by the Mees CCD (MCCD) imaging spectrograph. Although the BCS is not an imaging instrument, by comparing the temperature and emission measure determined from SXT filter ratios with the total of the stationary and blueshifted components observed by the BCS, Wulser et al. concluded that that the emissions in both instruments were coming from the same plasma. Since the footpoints seen in Hα were found to be co-spatial with those of the soft X-ray loop that spanned the neutral line, they also concluded that the bulk of the downward moving chromospheric plasma was located at the footpoint of the flare loop which contained the upward moving flare plasma, thus confirming the predictions of the chromospheric evaporation model. Having established that the site of the motions were co-spatial, they went on to determine that the total Hα downflow momentum was 3.6×10^{21} gm cm s^{-1}, while the upflow momentum in Ca XIX was 2.4×10^{21} gm cm s^{-1}. Within any errors they could calculate, they concluded that the two momenta were not significantly different.

They test the chromospheric evaporation model further by examining the heating in the chromosphere. Evaporation is expected to be most evident where massive heating is present; mechanism proposed for this heating include heating by non-thermal electrons and by conduction. By investigating the properties of the three kernels, they found that there was not enough power in the non-thermal electrons to create the observed downflows in the two Hα kernels associated with the flare loop – indeed, no significant hard X-ray emission is found near the kernel that showed strongest Hα redshift. However, at the third kernel there was enough power in the electrons and this was coincidentally the site of the strongest hard X-ray emission was observed (although it was weakest in Hα).

They conclude that in this event chromospheric evaporation could not have been caused by non-thermal electron heating associated with particle precipitation. But, from the dimensions and properties of the loop, they conclude that conductive heating is enough to explain the observed upflow velocities.

4.2   Event of 1991, December 16

The flare of 1991, December 16 was studied by Culhane et al. (1994).

The flare occurred at 04:56 UT on December 16, 1991 in NOAA AR 6961, located at N04 W45. Although this M2.7 flare was well observed by the *Yohkoh* HXT and BCS instruments, but because of the occurrence of a smaller (C7.4) flare in a different active region just prior to the flare being studied, the SXT instrument was centred elsewhere and only provides pre- and post-flare images. In the highest energy channel of HXT (HI; 53-93 keV) a simple hard X-ray burst was registered with a total duration of ~30 s; the simple nature of the burst was also emphasized by the 17 GHz lightcurves obtained by the Nobeyama Radio Observatory. A more gradual signal was recorded in the lower HXT channels and the BCS Ca XIX channel showed the gradual rise and decay that are characteristic of the heating and cooling of the flare plasma.

The flare morphology was studied with the Hard X-ray Telescope. Images for the HI channel showed two footpoints separated by 28 arcseconds (22,000 km); from this Culhane et al. deduced a loop length of 3.5×10^9 cm; with the cross-sectional area measure to be 1.0×10^{17} cm^2, they found a total loop volume of 3.5×10^{26} cm^3. The footpoints were active during the simple burst and the lightcurves for each footpoint showed a similar variation with time. In the LO channel (14-23 keV), the structure was more elongated with emission confined initially to the area of the footpoints, then moving out along an arc from these during the more gradual phase of the flare. The footpoints seen by HXT match the locations of flare emission points seen in Hα. However, in this case a momentum balance calculation was not possible since Hα line profiles were not observed at the loop footpoints. Pre-flare soft X-ray images show bright points that also match those seen in HXT and Hα. Overall, the pre- and post flare images from the SXT reveal little change in the structure of the active region.

Hard X-ray burst observed in the HXT HI channel are usually identified with the hard X-ray bremsstrahlung radiation resulting from the impact of non-thermal electrons in the chromosphere at the footpoints of the flare loop. For this event, the close synchronization of the lightcurves for each loop footpoints in HXT provides strong evidence that the structure is a simple loop and that the energetic electrons are created near the top of the loop and accelerated simultaneously towards the footpoints on either side to produce hard X-ray emission by the thick-target process.

As before, the velocity of the blueshifted component is derived from the BCS Ca XIX spectra (see Fig 3.). The hard X-ray burst corresponded to the time that the highest velocity blueshifts were seen in the Ca XIX channel; these corresponded to directed plasma flows of 800-900 km s^{-1}. This velocity is exceptional and represents one of the largest observed by the *Yohkoh* BCS. In order to ensure its validity, Culhane et al. had to make a very careful allowance for the presence of the flare that distracted the SXT (because this was at a different location, S12 E70, the spectrum from that event was shifted relative to that from the flare under study). Since the other event was well into its decay phase by the time the main event started, Culhane et al. believe that they were able to make a satisfactory correction for its emission.

The BCS also observed spectra in its Fe XXVI channel, suggesting the presence of a very hot component. From the temperatures derived for the loop top and footpoints, their temporal evolution, and the small emission measure of the hot plasma, Culhane et al. believe that the high temperature component of the plasma is concentrated at the footpoints rather than the loop top. They suggest that the "superhot"-like component in this event (and possibly others) is the result of violent heating of the chromosphere at the footpoints, rather than being a signature of the primary energy releases site as would be expected in the Type A events of Tanaka (1985). They also suggest that the factor of two longer lifetime of the "superhot" component for this flare may cause a reduction of the conductive heat flux by turbulence. However, the presence of turbulence cannot be confirmed in the BCS Fe XXVI data.

Culhane et al. went on to consider the energy balance of the event. From the hard X-ray data they determined that the total energy deposited by the non-thermal electrons during the burst is 1.0×10^{30} ergs. During the soft X-rays impulsive phase, from the parameters determined by the BCS, the thermal energy contained within the stationary component, including the conductive and radiative losses, totalled 1.1×10^{30} ergs; in the upflowing plasma the total was 0.3×10^{30} ergs (in both cases conductive losses were found to dominate). The total thermal energy in soft X-rays at the end of the evaporation was thus determined to be 1.4×10^{30} ergs. They suggest that the agreement of the energy in soft X-rays with the energy supplied by the non-thermal electrons indicates that the electron beam energy could have driven the plasma evaporation from the chromosphere.

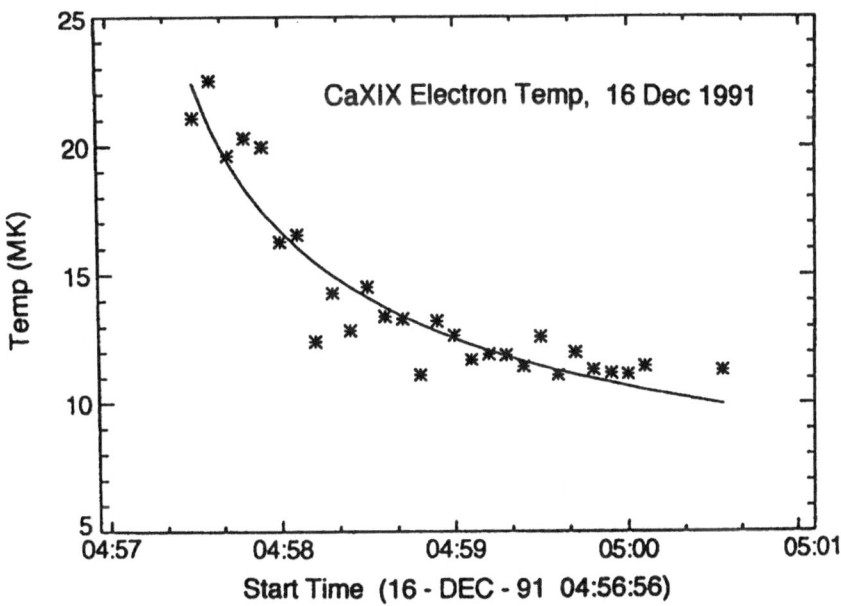

Fig. 6. Conductive cooling curve (solid line) superimposed on the fitted electron temperatures from the *Yohkoh* BCS Ca XIX channel for the flare of 1991, December 16 (after Culhane et al., 1994).

In this event, thermal conduction is much more effective (by a factor 10) than radiation as a cooling mechanism; this is related to the parameters of the loop. It is possible to investigate this by examining the conductive cooling time determined by the BCS Ca XIX data (see Fig 6). Using the loop dimensions determined from hard X-ray data, and applying the approach described by Culhane et al. (1970), a density $n_e = 2.5 \times 10^{11}$ cm^{-3} is implied. The density derived from observations by the BCS was $n_e = 3.6 \times 10^{11}$ cm^{-3}. These are in reasonable agreement and this implies that for this flare the cooling mechanism is by thermal conduction with a loop filling factor of the order unity.

4.3   Event of 1992, September 6

Electric currents and their associated electric fields have been shown to produce a viable mechanism for heating plasma and accelerating particles (e.g. Tsuneta, 1985) and can thus potentially explain both thermal and non-thermal processes in a flare. Zarro et al. (1995) test their DC-electric field model by applying it to the analysis of the flare of 1992, September 6.

The flare occurred at 08:57 UT on September 6, 1992 in NOAA AR 7270, located at S11 W38 - it was one of series of flares in that active region that had been well observed by Yohkoh. Zarro et al. used hard X-ray observations from the BATSE instrument (Fishman et al., 1989) on the *Compton Gamma-Ray Observatory* and the *Yohkoh* HXT, and soft X-ray observations from the *Yohkoh* BCS and SXT.

The flare geometry was determined using the SXT and HXT data: HXT showed that the separation of the flare footpoints was 28".5 and assuming a semicircular geometry, they deduced a loop length of 3.2×10^9 cm; they determined the loop cross-sectional area from soft X-ray images to be 2.0×10^{17} cm^2, giving a total loop volume of 1.3×10^{27} cm^3.

The flare energetics were determined from the BCS data. The Ca XIX spectra for this flare were exceptional because a strong blue asymmetry was evident before the peak in hard X-rays[2]. A secondary peak in the velocity of the blue component occurs at almost the same time as a brightening observed in SXT in one of legs of the loop; this they believe is consistent with a gradual filling of the loop with plasma. They determined the thermal heating rate needed to sustain the observed soft X-ray emission, finding that the conductive term dominates over the radiative term.

Zarro et al. (1995) consider that the presence of a dominant stationary component prior to the onset of the flare is not consistent with the thick-target electron beam model unless some additional form of preflare heating is invoked. The cause of such preflare heating is controversial and various explanations have been postulated. Zarro et al. suggest that it is produced by Joule heating (Ohmic dissipation) in thin current channels (or sheets) that are aligned parallel to the magnetic field. The channels they believe

---

[2] Note: A careful examination of the light-curves in Zarro et al. suggests that the initial peak of the upflow velocity, 270 km s^{-1} at ~09:02 UT, was at a very similar time to a weaker peak in hard X-rays which occurred before the main flare. This is best seen in the Log plot of the hard X-ray data - see Fig 7a.

generate an associated DC-electric field that is responsible for accelerating the electrons that produce the impulsive hard X-rays.

Zarro et al. parameterized the relative importance of acceleration versus heating by the ratio of the electron acceleration rate to the total Joule heating $(\dot{N}/Q)$ – this is dependent on electron temperature and density (T_e and n_e, these are determined from BCS data) and the electric field strength (E). This can be equated to ratio of the hard X-ray emission to the thermal heating rate implied by the soft X-ray observations (F_{HXR}/Q_{obs}) through the simple relationship:

$$F_{HXR}/Q_{obs} = \alpha \dot{N}/Q$$

where α is a constant of proportionality.

Prior to the impulsive phase the ratio of the DC-electric field to Dreicer field (the field at which runaway takes place) is sufficiently low that electron runaway acceleration is negligible compared to Joule heating. Consequently the thermal soft X-ray emission dominates over the non-thermal hard X-ray emission. The Joule heating raises the temperature at the loop top producing the dominant stationary component (and the soft X-ray enhancement at the loop top). Thermal conduction from the loop top then transports energy to the footpoints where it is converted to mass motions and produces the soft X-ray Ca XIX blueshifts.

Thermal conduction from the Joule-heated loop plasma continues to drive chromospheric evaporation and as the density of the loop increases, the non-thermal hard X-ray emission is quenched since the higher density requires a higher electric field before runaway can take place (Fig 7). An increase in E, triggered by reconnection between two intersecting loops, raises the Joule heating-rate thereby increasing the loop temperature and shifting a greater population of the thermal electrons closer to the threshold for runaway acceleration. Consequently, an impulsive increase in the non-thermal hard X-ray emission is produced.

Zarro et al. (1995) conclude that relative behaviours of the hard and soft X-rays can be understood in terms of simultaneous heating and acceleration in multiple current channels and that through their application of the DC-electric field model, physical parameters can be determined from the strength and temporal variations of the electric field in these channels.

The DC-electric field model of Zarro et al. has been tested on other flares (Zarro 1996) and again produces results that are consistent with the timing and energetics at the onset of the flares. The model is a useful tool for understanding the initial phases of the flare although it is less applicable later in the flares as other processes predominate.

5   Discussion and Conclusions

We have presented an overview of phenomena that can be observed in the corona that indicate energy is being released. The phenomena included range

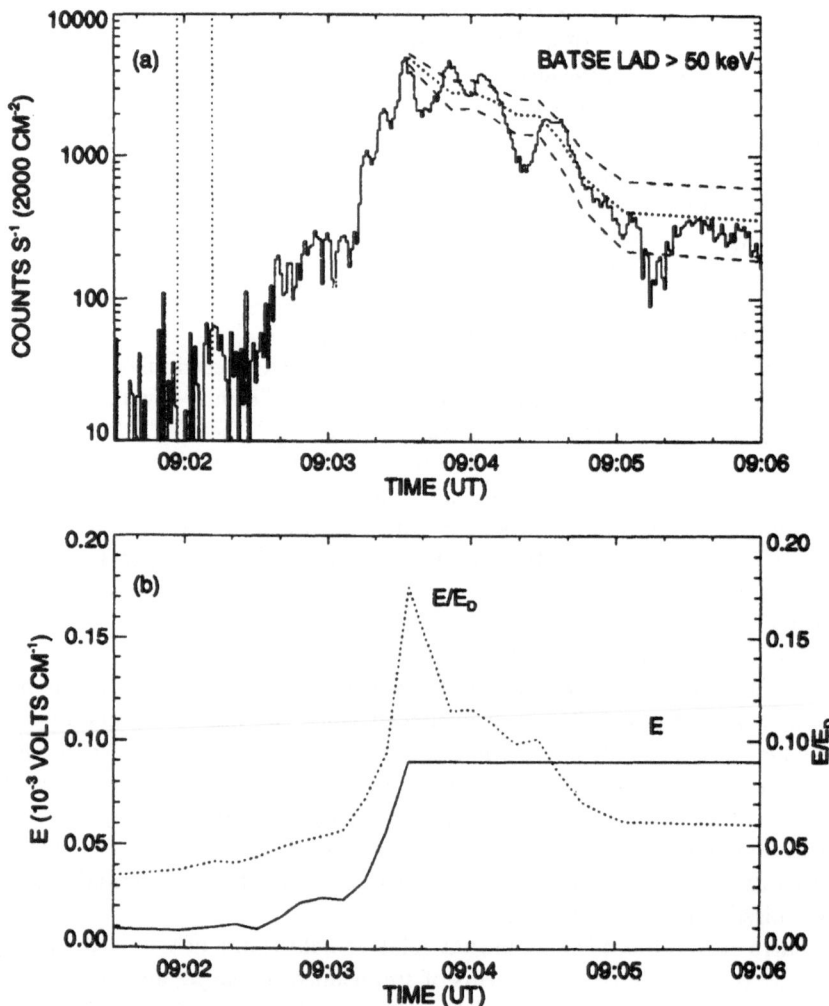

Fig. 7. The event of 1992, September 6. (a) The total BATSE hard X-ray (back-ground-subtracted) countrate above 50 keV. The dotted line following the curve shows the least-squares best-fit hard X-ray intensity F_{HXR} predicted by the DC-electric field model for a constant field strength $E = (9 \pm 1) \times 10^{-5}$ V cm^{-1}, where the uncertainty corresponds to the 1σ values. The vertical dotted lines near 09:02 UT mark the integration which corresponds to the maximum blueshifted velocity observed in BCS Ca XIX data; (b) The time variation of the best-fit electric field E (solid line) and the ratio $\epsilon = E/E_D$ (dotted line), where E_D is the Dreicer field. Note that E is assumed to be constant after the first hard X-ray peak (after Zarro et al., 1995).

from the smallest events that can be detected in X-rays to quite large events. We will now try to summarize some issues.

Recent observations suggest that the phenomena described in Section 2 (X-ray bright points, transient brightenings and jets) represent an extension at lower energies of flare events rather than being an entirely a new category of events. This is certainly implied by the soft X-ray data, although so far hard X-rays have not been recorded for these events. Many researchers think that this could be because the hard X-ray bursts fall below the sensitivity threshold of existing instruments and are thus not observed. The presence of accelerated non-thermal electrons is confirmed by the recent detection of Type III radio bursts in some of the events, and so it would appear that the required mechanisms to produce hard X-rays are present. Perhaps future advances in instrumentation will confirm the relationship between these events and flares. If so, then the same basic mechanism (probably reconnection) could be responsible for the very smallest to the largest forms of energy release in the corona.

In Section 3, the bluewing and non-thermal broadening seen in some flares were detailed. The spatial and temporal relationships between the hard X-ray burst and soft X-ray emission are crucial to understanding the mechanisms responsible for flares. Some of the confusion that currently exists over which comes first, the bluewing or hard X-ray burst, may (again) be related to the sensitivity of current instruments and this will hopefully be resolved shortly. The parameters described provide constraints for any models since they give us clues of where, when and how the energy is released. Any model must be able to explain the phenomena seen in both hard and soft X-rays, otherwise it has to be missing some fundamental details of the energy release.

In the three events discussed in detail in Section 4, all exhibited the signature of chromospheric evaporation, a clear indicator of energy release, but the way in which energy is transported to the chromosphere was found to differ. In many flares, a particle beam impacting on the chromosphere provides the source of energy, but in some conduction from the primary release site high in the corona drives the evaporation. For the events of November 15, 1991 and September 6, 1992 conduction was thought to have heated the chromosphere and evaporated the material, but for the event of December 16, 1991 the precipitation of energetic particles appears to be the cause of the evaporation. The DC-electric field model discussed by Zarro et al. (1995) is one of many models that cover the onset of a flare. The model is able to describe many aspects of the impulsive phase of the flare studied but how generally applicable the model is is not clear and more work is needed.

For any event, the way the flare evolves clearly has some dependency on its size and morphological complexity. The availability of energy, and mechanisms by which it can be released are also important. The preflare configuration of the flare site can determine how the energy release of the flare subsequently develops, but the complexity of the structures involved make a full description difficult. It is possible to quantify smaller, more numerous energy releases in scaling laws, but these fall down for large events where many processes can be acting simultaneously.

All events are not the same and trying to use individual events to justify particular theories is fraught with problems. For almost any event, many models apply - what is key is why a particular model is favoured in a particular circumstance, and why certain models are not applicable in some events. If we could understand this, then we would be well on the way to understanding some basic principles of how the Sun works.

References

Alexander A. (1990): Soft X-ray Line Broadening in Solar Flares, A&A, **236**, L9.

Antonucci E., Gabriel A.H., Acton L.W., Culhane J.L., Doyle J.G., Leibacher J.W., Machado M.E., Orwig L.E., Rapley C.G. (1982): Impulsive Phase of Flares in Soft X-ray Emission, Solar Phys., **78**, 107.

Antonucci E., Dodero M.A., Martin R. (1990): High-velocity Evaporation during the Impulsive Phase of the 1984 April 24 Flare, ApJ (Suppl.), **73**, 137.

Bely-Dubau F., Faucher P., Steenman-Clarke L., Dubau J., Loulergue M., Gabriel A.H., Antonucci E., Volente S., Rapley C.G. (1982): Dielectronic Satellite Spectra for Highly Charged Helium-like Ions. VII - Calcium spectra: Theory and Comparison with SMM observations, MNRAS, **201**, 1155.

Bentley R.D., Lemen J.R., Phillips K.J.H., Culhane J.L. (1986): Soft X-ray Observations of High-velocity Features in the 29 June 1980 Flares, A&A, **154**, 255.

Bentley R.D. (1994): The Relationship between the Soft X-ray Blue Wing and the Hard X-ray Burst, *X-ray Solar Physics from Yohkoh*, eds. Uchida Y., Hudson H.S., Watanabe T., Shibata K., University Academy Press, Inc., 87.

Bentley R.D., Doschek G.A., Simnett G.M., Rilee M.L., Mariska J.T., Culhane J.L., Kosugi T., Watanabe T. (1994): The Correlation of Solar Flare Hard X-ray Bursts with Doppler Blueshifted Soft X-ray Flare Emission, ApJ (Letters), **421**, L55.

Bentley R.D. (1996): Mass Motions in Flares, *Magnetodynamic Phenomena in the Solar Atmosphere - Prototypes of Steller Magnetic Activity*, eds. Uchida Y., Kosugi T., Hudson H.S., Kluwer Academic Publishers, 177.

Canfield R.C., Zarro D.M., Metcalf T.R., Lemen J.R. (1990): Momentum Balance of Four Solar Flares, ApJ, **348**, 333.

Canfield R.C., Hudson H.S., Leka K.D., Mickey D.L., Metcalf T.R., Wuelser J.-P., Acton L.W., Strong K.T., Kosugi T., Sakao T. (1992): The X-flare of 1991 November 15 - Coordinated Mees/Yohkoh Observations, PASJ, **44**, L111.

Cheng C.-C., Rilee M., Uchida Y. (1994): Thermal and Nonthermal Energizations in Solar Flares: Soft X-ray Spectroscopic and Hard X-ray Observations, *Proc. of Kofu Symposium*, eds. Enome S., Hirayama T., **NRO Report No. 360**, 213.

Crosby N.B., Aschwanden M.J., Dennis B.R. (1993): Frequency Distributions and Correlations of Solar X-ray Flare Parameters, Solar Phys., **143**, 275.

Culhane J.L., Vesecky J.F., Phillips K.J.H (1970): The Cooling of Flare Produced Plasma in the Solar Corona, Solar Phys., **15**, 394.

Culhane J.L., Hiei E., Doschek G.A., Cruise A.M., Ogawara Y., Uchida Y., Bentley R.D., Brown C.M., Lang J., Watanabe T., Bowles J.A., Deslattes R.D., Feldman U., Fludra A., Guttridge P., Henins A., Lapington J., Magraw J., Mariska J.T., Payne J., Phillips K.J.H, Sheather P., Slater K., Tanaka K., Towndrow E., Trow M.W., Yamaguchi A. (1991): The Bragg Crystal Spectrometer for SOLAR-A, Solar Phys., **136**, 89.

Culhane J.L., Phillips A.T., Inda-Koide M., Kosugi K., Fludra A., Kurokawa H., Makishima K., Pike C.D., Sakao T., Sakuai T., Doschek G.A., Bentley R.D. (1994): Yohkoh Observations of the Creation of High-Temperature Plasma in the Flare of 16 December 1991, Solar Phys., **153**, 307.

Dennis B.R. (1985): Solar Hard X-ray Bursts, Solar Phys., **100**, 465.

Doschek G.A., Strong K.T., Bentley R.D., Brown C.M., Culhane J.L., Fludra A., Hiei E., Lang J., Mariska J.T., Phillips K.J.H., Pike C.D., Sterling A.C., Watanabe T., Acton L.W., Bruner M.E., Hirayama T., Tsuneta S., Rolli E., Kosugi T., Yoshimori M., Hudson H.S., Metcalf T.R., Wuelser J.-P., Uchida Y., Ogawara Y. (1993): The 1992 January 5 Flare at 13.3 UT: Observations from Yohkoh, ApJ, **416**, 845.

Feldman W.C., Gosling J.T., McComas D.J., Phillips J.L. (1993): Evidence for Ion Jets in the High-Speed Solar Wind, JGR, **98**, 5593.

Fishman G.J., Meegan C.A., Wilson R.B., Parnell T.A., Paciesas W.S., Pendleton G.N., Hudson H.S., Matteson J.L., Peterson L.E., Cline T.L (1989): The BATSE Experiment on the Gamma Ray Observatory: Solar Flare Hard X-ray and Gamma-ray Capabilities, *Max '91 Workshop 2: Developments in Observations and Theory for Solar Cycle 22*, eds. Winglee R.M., Dennis B.R., 96.

Fludra A.F., Lemen J.R., Jakimec J., Bentley R.D., Sylwester J. (1989): Turbulent and Directed Plasma Motions in Solar Flares, ApJ, **344**, 991.

Golub L., Krieger A.S., Viana G.S. (1976): Distribution of Lifetimes for Coronal Soft X-ray Bright Points, Solar Phys., **49**, 79.

Harvey K.L. (1985): The relationship between coronal bright points as seen in He I Lambda 10830 and the evolution of the photospheric network magnetic fields, Aust. J. Phys., **38**, 875.

Harvey K.L. (1996): Observations of X-ray Bright Points, *Magnetic Reconnection in the Solar Atmosphere*, eds. Bentley R.D., Mariska J.T., Astronomical Society of the Pacific Conference Proceedings, 9.

Khan J.I., Harra-Murnion L.K., Hudson H.S., Lemen J.R., Sterling A.C. (1995): Yohkoh Soft X-ray Spectrocsopic Observtions of the Bright Loop-top Kernals of Solar Flares, ApJ (Letters), **452**, L153.

Kosugi T., Makishima K., Murakami T., Sakao T., Dotani T., Inda M., Kai K., Masuda S., Nakajima H., Ogawara Y., Sawa M., Sibasaki K. (1991): The Hard X-Ray Telescope (HXT) for the SOLAR-A Mission, Solar Phys., **136**, 17.

Kundu M.R., Strong K.T., Pick M., White S.M., Hudson H.S., Harvey K.L., Kane S.R. (1994): Nonthermal processes in flaring X-ray-bright points, ApJ, **427**, L59.

Kundu M.R., Raulin J.P., Nitta N., Hudson H.S., Shimojo M., Shibata K., Raoult A. (1995): Detection of Nonthermal Radio Emission from Coronal X-Ray Jets, ApJ, **447**, L135.

Mariska J.T., Doschek G.A., Bentley R.D. (1993): Flare plasma dynamics observed with the Yohkoh Braagg Crystal Spectrometer, I. Properties of the Ca XIX Resonance Line, ApJ, **419**, 418.

Mariska J.T., Sakao T., Bentley R.D. (1995): Hard and Soft X-ray Observations of Solar Limb Flares, ApJ, **459**, 815.

Masuda S. (1993): Hard X-ray Sources and the Primary Energy Release Site in Solar Flares, *Thesis dissertation*, School of Science, University of Tokyo.

Masuda S., Kosugi T., Hara H., Tsuneta S., Ogawara Y. (1994): A Loop-top Hard X-ray Source in a Compact Solar Flare as Evidence for Magnetic Reconnection, Nature, **371**, 495.

Plunkett S.P., Simnett G.M. (1994): Temporal Correlation of Solar Hard X-ray Bursts with Chromospeheric Evaporation, Solar Phys., **155**, 351.

Saba J.L.R., Strong K.T. (1986): Evidence for Coronal Turbulence in a Quiescent Active Region, Adv. Space Res., **6**, 37.

Sakao T., Kosugi T., Masuda S., Inda M., Makishima K., Canfield R.C., Hudson H.S., Metcalf T.R., Wuelser J.-P., Acton L.W. (1992): Hard X-ray Imaging Observations by Yohkoh of the 1991 November 15 Solar Flare, PASJ, **44**, L83.

Savy S.K. (1994): Private Communication.

Shibata K., Ishido Y., Acton L.W., Strong K.T., Hirayama T., Uchida Y., McAllister A.H., Matsumoto R., Tsuneta S., Shimizu T. (1992): Observations of X-ray jets with the Yohkoh Soft X-ray Telescope, PASJ, **44**, L173.

Shibata K., Shimojo T., Yokoyama T., Ohyama M. (1996): Theory and Observations of X-ray Jets, *Magnetic Reconnection in the Solar Atmosphere*, eds. Bentley R.D., Mariska J.T., Astronomical Society of the Pacific Conference Proceedings, 29.

Shimizu T., Tsuneta S., Acton L.W., Lemen J.R., Uchida Y. (1992): Transient Brightenings in Active Regions Observed by the Soft X-ray Telescope on Yohkoh, PASJ, **44**, L147.

Shimizu T., Tsuneta S., Acton L.W., Lemen J.R., Ogawara Y., Uchida Y. (1994): Morphology of Active Region Transient Brightenings with the Yohkoh Soft X-ray Telescope, ApJ, **422**, 906.

Shimizu T. (1995): Energetics and Occurrence Rate of Active-Region Transient Brightenings and Implications for the Heating of the Active-Region Corona, PASJ, **47**, 251.

Shimizu T. (1996): Yohkoh Observations Related to Coronal Heating, *Magnetic Reconnection in the Solar Atmosphere*, eds. Bentley R.D., Mariska J.T., Astronomical Society of the Pacific Conference Proceedings, 59.

Shimojo M., Hashimoto S., Shibata K., Hirayama T., Hudson H.S., Acton L.W. (1996): Statistical Study of Solar X-Ray Jets Observed with the Yohkoh Soft X-Ray Telescope, PASJ, **48**, 123.

Strong K.T., Hudson H.S., Dennis B. (1994): Evidence for Impulsive Soft X-ray Bursts during Flares, *X-ray Solar Physics from Yohkoh*, eds. Uchida Y., Hudson H.S., Watanabe T., Shibata K., University Academy Press, Inc., 65.

Tanaka K., Watanabe T., Nishi K., Akita K. (1982): High-resolution Solar Flare X-ray Spectra obtained with Rotating Spectrometers on the Hinotori Satellite, ApJ (Letters), **254**, L59.

Tanaka K. (1985): Impact of X-ray Observations from the Hinitori Satellite on Solar Flare Research, PASJ, **39**, 1.

Tsuneta S. (1985): Heating and acceleration processes in hot thermal and impulsive solar flares, ApJ, **290**, 353.

Tsuneta S., Acton L., Bruner M., Lemen J., Brown W., Carvelho R., Catura R., Freeland S., Jurcevich B., Morrison M., Ogawara Y., Hirayama T., Owens J. (1991): The Soft X-Ray Telescope for the SOLAR-A Mission, Solar Phys., **136**, 37.

Winglee R.M., Kiplinger A.L., Zarro D.M., Dulk G.A., Lemen J.R. (1991): Interpretation of Soft and Hard X-ray Emissions during Solar Flares. I. Observations, ApJ, **375**, 366.

Wulser J.-P., Canfield R.C., Acton L.W., Culhane J.L., Phillips A.T., Fludra A., Sakao T., Masuda S., Kosugi T., Tsuneta S. (1994): Multispectral Observations of Chromospheric Evaporation in the 1991 November 15 X-class Solar Flare, ApJ, **424**, 459.

Zarro D.M. (1993): Flare Dynamics Observed in S XV, *UV and X-ray of Laboratory and Astrophysical Plasmas*, eds. Silver E., Kahn S., Cambridge University Press, 603.

Zarro D.M., Mariska J.T., Dennis B.R. (1995): Testing the DC-Electic Field Model in a Solar Flare Observed by Yohkoh and the Compton Gamma-Ray Observatory, ApJ, **480**, 888.

Zarro D.M. (1996): Energetic Consequnces of the DC-Electric Field Model, *Magnetic Reconnection in the Solar Atmosphere*, eds. Bentley R.D., Mariska J.T., Astronomical Society of the Pacific Conference Proceedings, 209.

Solar Flare Radio and Hard X-Ray Observations and the Avalanche Model

Nicole Vilmer and Gérard Trottet

LPSH - DASOP (URA 2080), Observatoire de Paris, Section d'Astrophysique de Meudon, F-92195 Meudon Cedex, France

Abstract. This paper reviews hard X-ray and radio observations obtained with high temporal and/or spatial resolution which show that the production of non thermal electrons is fragmented in both space and time. Observations dealing with the extent in heights and spatial scales of the electron acceleration sites are also reviewed. The knowledge of the electron energy spectra deduced at low and high energies from both hard X-ray and centimeter/millimeter observations is discussed. The experimental data is compared with the predictions of the avalanche model for solar flares and further tests of the model proposed.

Résumé. Ce papier présente une revue des observations X dur et radioélectriques obtenues à haute résolution temporelle et/ou spatiale qui montrent l'aspect fragmenté tant dans l'espace que dans le temps de la production des électrons non thermiques. Certaines observations concernant l'extension en altitudes et en échelles spatiales des sites d'accélération des électrons sont également présentées. Notre connaissance des spectres en énergie des électrons déduits à basse et haute énergie des observations X dur et centimétriques/millimétriques est discutée. Les observations sont enfin comparées aux prédictions du modèle d'avalanche des éruptions solaires et des tests complémentaires de ce modèle sont proposeés.

1 Introduction

Non-thermal particles play a major role during flares since they contain a large amount of the released energy. Flare particle acceleration occurs on time scales ranging from a few seconds or less to hours. It has now been studied for many decades using observational diagnostics provided by the radiation from the energetic particles when they interact with the solar atmosphere and direct measurements of solar energetic particles in the interplanetary space, as well as through models for energy release and acceleration processes. The most direct diagnostics of energetic electrons from a few tens to a few hundreds keV is the X- ray bremsstrahlung that they produce in the denser and lower regions (chromosphere and low corona) of the solar atmosphere. For a fraction of solar flares the bremsstrahlung continuum is found to extend above 10 MeV (see e.g. Vestrand et al. 1991, Vilmer et al. 1994), sometimes up to a few hundred of MeV (e.g. Akimov et al. 1991, Leikov et al. 1993; Kanbach et al. 1993), implying the acceleration of ultrarelativistic electrons.

Complementary observations of energetic electrons are provided by the gyrosynchrotron radio emission that they produce in the low corona in the centimeter/millimeter wavelengths. Finally, the coherent plasma radiations, such as the ones emitted in the meter/decimeter domains at greater heights (10^4 - 10^5 km above the photosphere) represent the most sensitive diagnostics of weakly energetic electrons (a few tens of keV). The observations of the whole radio spectrum allows us to probe the solar atmosphere on a wide range of heights going from the low corona to the interplanetary medium. It has been shown through the analysis of several events that only multiwavelengths studies provide a comprehensive approach of the general problem of particle acceleration and transport in solar flares (e.g. Trottet 1986). This review is focused on the only diagnostics of electron acceleration as deduced from radio and X-ray observations. An accompanying review on diagnostics of both electron and ion acceleration can be found in the Proceedings of the VIII European Solar Physics Meeting (Trottet and Vilmer 1996).

Two complementary approaches have been used to infer from the observations information on particle acceleration in solar flares. Many studies have been performed on detailed analysis of individual events, allowing to get constraints on e.g. the number and the energy spectrum of accelerated particles and the characteristics of the magnetic structures at different scales in which energetic particles are produced, propagate and radiate...Such an approach is essential to model the acceleration mechanisms at work in solar flares. It was already suggested by Kaufmann (1985) that solar events may result from the convolution of many fast primary energetic electrons at various repetition rates. This was followed in the last years by many statistical studies performed on the large hard X-ray data bases acquired on more than ten years (e.g. Crosby et al 1993, Lu et al 1993, Bromund et al 1995). The goal of such studies is to describe the global behavior of thousands of solar flares · by determining the occurrence distributions of flare parameters and to give some insight into the nature of the flaring system. Several "cartoons" have been developed to describe the context in which "non thermal" energy release and particle acceleration occur. Solar flares have thus been interpreted either as resulting from the evolution of very simple magnetic topologies such as magnetically reconnecting interacting loops or of a complex dynamic system implying the interaction of thousands of small scale simple structures (e.g. Vlahos 1994).

In the following we shall review observations of non-thermal radiation in flares obtained with high temporal and/or some spatial resolution dealing with the context of non-thermal energy release. These observations suggest a fragmentary nature of the energy release in time (Sect.2) and space (Sect.3). Section 4 will present recent results concerning the extent in height and spatial scale of the electron acceleration sites. In Sect.5, we shall address the problem of the determination of the flare accelerated energy spectrum for electrons from both bremsstrahlung and gyrosynchrotron radiation processes.

In Sect.6, we confront the avalanche model with these experimental facts and propose further tests of the model.

2 Temporal Fragmentation of Energy Release in Flares

2.1 Distribution of Fast Time Structures in Hard X-ray Bursts

The study of rapidly time varying emissions from energetic electrons has great potential to infer the time scales of non-thermal energy release in solar flares. Early studies performed on TD1A observations (Van Beek et al. 1974, de Jager and de Jonge 1978) had shown that hard X-ray flares can be decomposed in a series of short lived bursts of duration ranging from a few seconds to a few tens of seconds (" Elementary Flare Bursts"). With the constantly increasing sensitivity of hard X- ray experiments, analyses of faster and faster time variations have been systematically performed on the data (e.g. Kiplinger et al. 1983). It has now been found that time scales of a few hundreds of milliseconds are almost systematically observed in solar flares. It shows that the electrons produced in these elementary energy release phenomena must interact in dense regions ($\geq 10^{12}$ cm^{-3}) and indicates that the flare energy release is fragmented. While Vilmer et al. (1996) based their analysis on more than 100 bursts recorded by the PHEBUS experiment around 100 keV with a basic time accumulation $\leq$ 31.25 ms, Aschwanden et al. (1995a) analyzed more than 600 solar flares recorded with a time resolution of 64 ms in the 25-50 keV and 50-100 keV energy range with the BATSE/CGRO experiment. Although the two studies performed independently on the two data sets slightly differ in their definition and technique of analysis of time structures, they lead to consistent and complementary results:

1. A large percentage of bursts (more than 70 %) presents time features with rise times in the 100ms-1s range around 100 keV or time structures ("HXR pulses") with duration widths in the 300ms-3s range in the 25-100 keV range. As an example, Fig.1a (Vilmer et al. 1996) shows the time history of the 0.10-1.6 MeV count-rate observed by PHEBUS with an accumulation time of 93.75ms for an event where 40 time structures are found with rise times less than 1 second.
2. There is a continuous distribution of the values of the time scales found in a single burst as well as for all the bursts of the sample (Vilmer et al. 1996). The values of the rise times of the 40 structures found in the 14 March 1991 event (Fig.1a) span the whole time scale range between 93.25 ms and 968 ms, with more than 30 of the features being distributed below 500ms (Fig.1b). If a hard X-ray burst is characterized by the time scale (t_{cm}) for which the maximum number of significant features is detected, the occurrence distribution of the number of bursts as a function of t_{cm} is rather flat in the 100ms-900ms time range (Vilmer et al. 1996). This

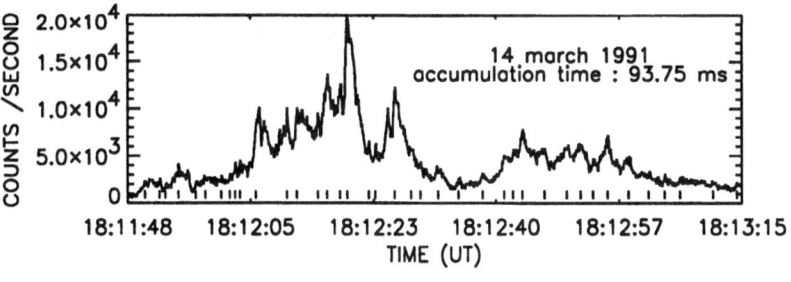

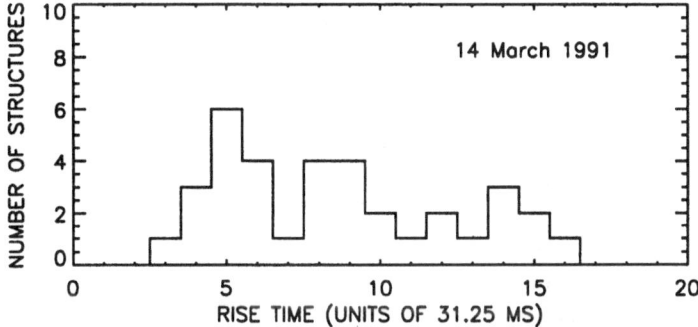

Fig. 1. (from Vilmer et al. 1996) Count-rate of the 14 March 1991 event in the 0.1- 1.6 MeV range. The dashes indicate the location of the time variations with significant rise times less than 1 second (a). Distribution of rise times below 500 ms for the 14 March 1991 event (b)

indicates that there is no characteristic time scale holding for the many bursts observed.

3. The occurrence distribution of bursts as a function of the shortest rise time t_{min} found around 100 keV can be represented by an exponential function in the 100ms-900ms range ($N(t_{min}) \simeq \exp(-t_{min}/0.47s)$) (Vilmer et al. 1996). Such a distribution is also found for the durations of the HXR pulses identified in the BATSE observations in both the 25-50 keV and 50-100 keV range with a time constant of 0.4s (Aschwanden et al. 1995a). The similarity between the distributions obtained in different energy ranges indicates that the pulse duration is more determined by the electron injection than by transport effects, so that it strongly supports the interpretation that the time scales derived from such analysis are a good indicator of the electron acceleration time scales. Under these conditions, the shape of the distribution of time structures as a function of their duration shows that non thermal energy release can be understood as a random process governed by Poisson statistics (Aschwanden et al. 1995a).

4. The distribution of the peak count-rate F above background of the HXR pulses can be described by a power-law $N(F) \simeq F^{-1.8}$ in the 50-100 keV range while in the lower energy band (25-50 keV) the distribution exhibits a larger deviation from a power law and a different slope which may be attributed to a larger thermal contamination (Fig.2) (Aschwanden et al. 1995a). It is worth noting that the size distribution of the HXR pulses above 50 keV has a similar slope as the size distribution of global HXR flares above 25 KeV (e.g. Crosby et al. 1993).

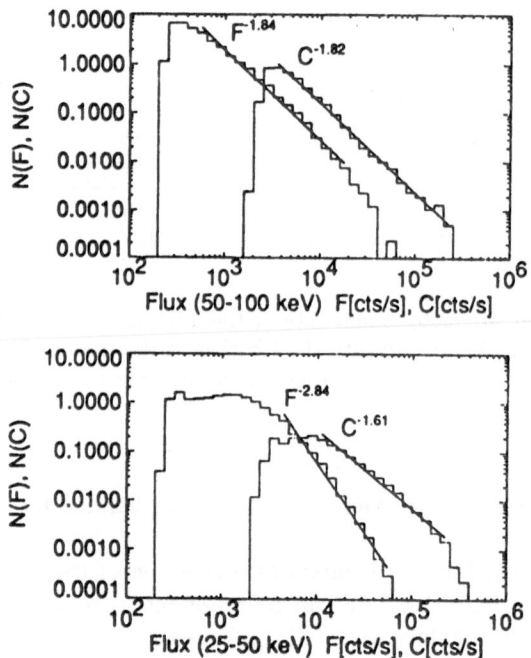

Fig. 2. Distribution of the peak count-rates F and C of the HXR pulses. The background-subtracted pulse count-rate is denoted by F, while C represents the total count-rate including background (from Aschwanden et al. 1995a)

2.2 Simultaneous Detection of Fast Time Structures in Hard X-rays and at Radio Wavelengths

The observations that even the simplest impulsive bursts at hard X-rays and microwaves present time scales in the range of 10-100ms led Kaufmann (1985) to suggest that the energy release rate is comparable in small events and large events. This may be explained if bursts at different wavelengths result from the convolution of many fast primary energetic injections, at different repetition rates (Kaufmann 1985). More recently, Aschwanden et al. (1995b) performed a statistical survey of energetic electron beams simultaneously producing HXR pulses and type III-like meter- decimeter radio bursts. The identification of simultaneous events is based on the detection of a normal or reverse type III burst coinciding (within 1s) with a HXR pulse of similar duration (1s). The occurrence distribution of the HXR pulse count-rate of the sample selected on this criterion exhibits the same behavior as the one obtained on the whole BATSE data set (Aschwanden et al. 1995a).This indicates that the selection does not introduce a bias on the hard X-ray pulses. However, the frequency distribution of the fluxes of the associated radio pulses is characterized by a power law distribution with a different slope of -1.3. The difference in the slope may be attributed to the fact that contrary to the X-ray emission, the production of type III-like radio emission is an incoherent process. The slope of the distribution is thus not linked in a straightforward way to the non-thermal energy release mechanism. On the other hand, as in the case of HXR pulses, the distribution of the values of the radio bursts duration (Fig.3) is also represented by an exponential function between 0.3 and 1.3s (Aschwanden et al 1995b).

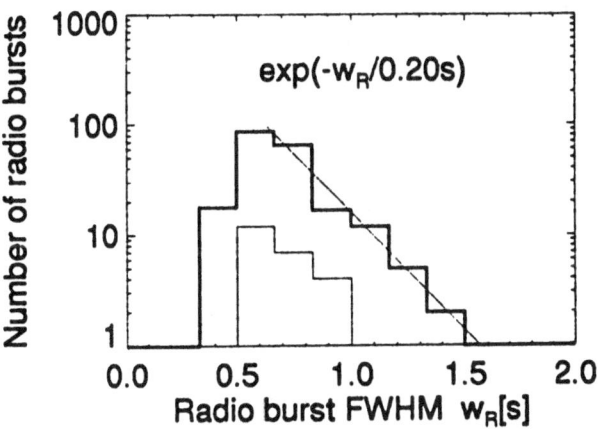

Fig. 3. Distribution of time scales (FWHM) of radio bursts observed with PHOENIX associated with simultaneous HXR pulses detected with HXRBS (thick lines) and BATSE (thin lines). The linear regression fits correspond to exponential distributions (adapted from Aschwanden et al. 1995b)

2.3   Time Structures in Electron-Dominated Events at Energies Above 10 MeV

SMM observations have shown in some events the occurrence in the course of a flare of peaks lasting a few seconds to a few tens of seconds during which the continuum bremsstrahlung emission extends above 10 MeV and dominates ("electron dominated events" or peaks) (Rieger and Marschhaüser 1990, Marschhaüser et al. 1994, Rieger 1994). A significant time structure of the order of 1s has also been reported at energies as high as 10 MeV by Pelaez et al. (1992) for one event in which the high energy emission is predominantly radiated by ultrarelativistic electrons. The detection of time structures of the order of the second up to 10 MeV energies confirms that electrons must interact in dense regions (see Sec.2.1). Furthermore, the observations of these short duration bremsstrahlung bursts in the course of a flare have been interpreted as suggesting the sudden appearance of transient potential drops in the solar atmosphere leading to a preferential acceleration of ultrarelativistic electrons (Chupp 1996) (see also Sec.3.2 and 5).

3   Evolution of the Spatial Configuration during Flares and Particle Acceleration

3.1   Fragmentation in Space and Time of the Production of Non-thermal Particles

Millimeter radio observations obtained at 48 GHz with the multi-beam antenna of the Itapetinga Radio Observatory which provide the centroïd position of bursts with a spatial accuracy of 5" to 20" and a time resolution down to a few milliseconds have clearly shown variations in space of the positions of emitting sources on time scales of a few seconds to a few tens of seconds (e.g. Herrmann et al. 1994, Correia et al. 1995). Rapid changes of the localization of time features of a few seconds duration are observed over a distance of the order of 10" during the development of a solar burst (Herrmann et al. 1994). Correia et al. (1995) show another example of the complex and fast evolution of the millimeter source during a short burst (60s). As seen on Fig.4a the burst intensity time profile at 48 GHz exhibits fast pulses of about a few seconds (capital letters) clustered in main bursts lasting 20 to 30 seconds (capital numbers). During bursts I,II and III the centroid of the emission arises from different locations. Furthermore, the spatial analysis of the fast pulses shows that each pulse originates from a distinct location separated in space by about 5" to 6" and stable during the pulse duration (see Fig. 4b from Correia et al. (1995) showing the dynamic map for time interval III). During pulse G (8s), the centroid position is always located in the upper left part of the pattern, while during pulse H (11s) it is observed in the lower right part of the pattern. These observations suggest that the impulsive solar burst is a superposition of small events spread both in time and space. The stability of the centroid position during the fast pulses A to G indicates that the sub-second fluctuations observed in the time profiles cannot be spatially resolved. They are produced in a small area of diameter of about 3" (near the spatial resolution of the instrument) covered by the dispersion of the centroid during the pulse. Fig. 4c shows a dynamic map of the centroid positions of the radio sources observed every second at 435 MHz with the Nançay Radioheliograph during the same time interval. The millimeter pulses G and H are associated with two positions separated at decimeter wavelengths by $0.1R_S$. A similar behavior is observed during peaks I and II at 435 MHz, where the emission arises from distinct sources lightening at different times and separated by tenths of solar radii. The interpretation of the link between the small scale magnetic structures revealed by millimeter wave emission and the larger scale magnetic structures traced by the decimeter emission is not straightforward. This observation suggests however that, even in the course of a short burst, different large scale as well as short scale magnetic structures are involved in the release and transport of energetic electrons.

The first VLA observations of a solar narrow-band millisecond spike event (Krucker et al. 1995) characterized with the PHOENIX spectrometer exhibit similar variations in space and time of the radio emission sources. Within 50 seconds, about 50 single spikes are observed. The VLA observations obtained with a time resolution of 1.6 s allow to probe the spatial structure of the emission. However, as the VLA time resolution is much larger than spikes duration (typically 100ms), there will often be more than one spike in an image so that it is not possible to observe changes of positions from one spike to the other but rather from groups of spikes to others. The major spikes observed by PHOENIX arise from at least 3 locations separated by up to 130" exhibiting different time profiles (Fig. 5). Emission is observed in source B for all the strongest spikes, being predominant for the first group. For periods 2,3 and 4, the emission comes predominantly from either C or A. Furthermore, as shown in Fig.5c which represents the coordinates of the local emission maxima as a function of time, there is a shift of the positions of the main components A, B, C which may indicate that the emission is produced in different unresolved structures at different times. The production of non-thermal electrons thus originates in a complex environment.

3.2   Evolution of the Characteristics of the Energetic Particles and of the Spatial Configuration

Several multiwavelength studies have been performed combining hard X-ray/γ-ray spectral measurements with meter/decimeter radio imaging observations (e.g. Klein et al. 1983, Raoult et al. 1995, Chupp et al. 1993, Trottet et al 1993, Trottet et al, 1994). They indicate that hard X-ray and radio emissions arise from electrons produced by a common acceleration/injection (e.g. Trottet 1986 for a review). The observations show that the different elementary X-ray/γ-ray peaks are associated with different radio sources. Because

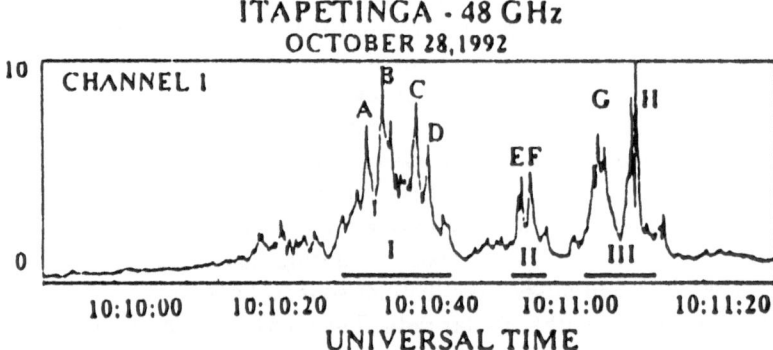

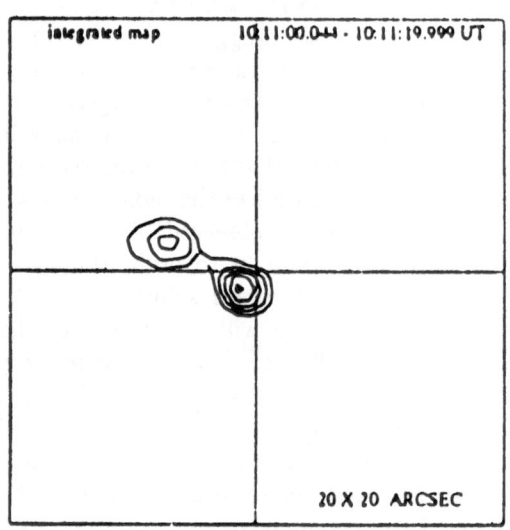

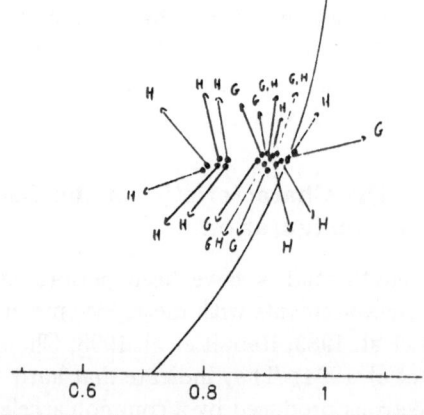

Fig. 4. Time profile of the October 28, 1992 burst at 48 GHz (a) (Correia et al. 1995). Dynamic burst maps at 48 GHz for the time interval III (b) (Correia et al. 1995). Dynamic map of the centroid positions of the radio sources observed every second at 435 MHz with the Nançay Radioheliograph. The letters indicate the time (during millimeter pulse G or H) when the source is observed (c).

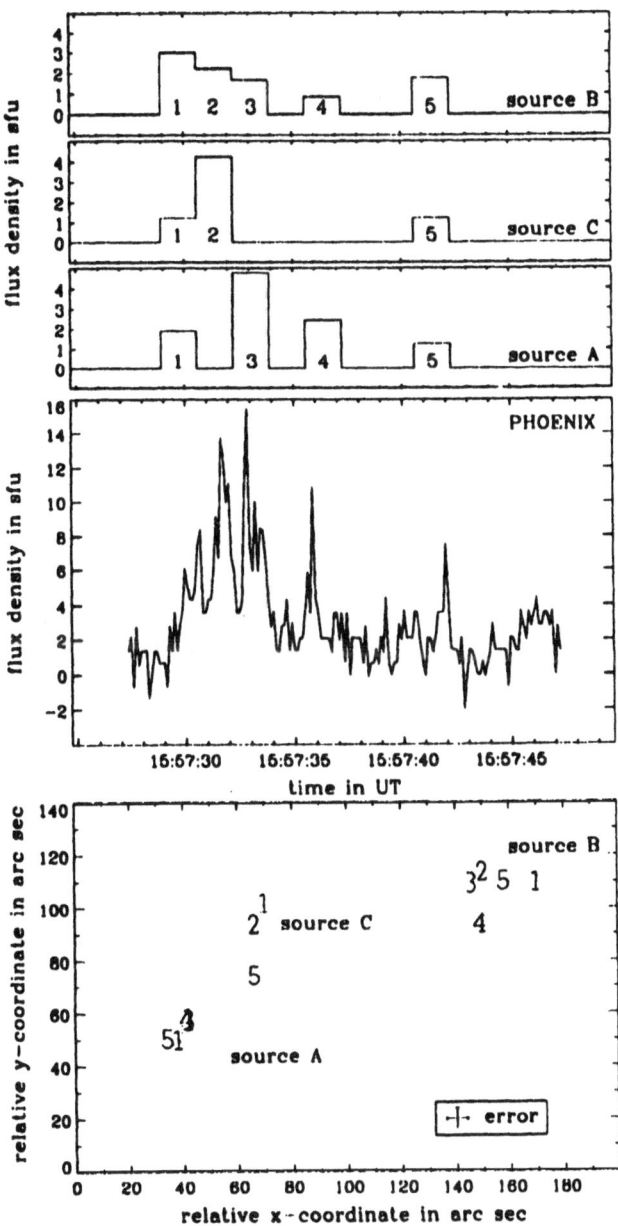

Fig. 5. (from Krucker et al. 1995) Top Time profiles of the flux density of the three spatial components observed with the VLA during the 7 August 1989 narrow-band millisecond spike event. Middle Time profile of the flux density of the event observed at 333 MHz with a time integration of 0.12s with the PHOENIX spectrometer. Bottom Time evolution of the source positions observed with the VLA. The numbers 1 to 5 indicate the strongest spikes observed with PHOENIX.

of the short delay between the electron production (as revealed by X-ray observations) and the changes in the radio emission pattern, the evolution of the localization of the large scale emitting sources has been attributed to rapid changes of the smaller scale magnetic structures involved in the acceleration (see e.g. Trottet, 1994 for a review). Noticeable changes in the accelerated electron spectrum and in the ratio of accelerated electrons and ions are observed in close connection with the changes of the spatial distribution of the electron emitting sites in the large scale magnetic structures (Chupp et al. 1993, Trottet et al. 1994). The production of very energetic electrons revealed by a high energy bremsstrahlung component may also be associated with new remote Hα bright features appearing at the border of the main flare site. Such a behavior has been reported for high energy peaks (photon emission above 10 MeV) occurring in the development of two hard X-ray/γ-ray events associated with Hα disk flares (Wülzer et al. 1990, Chupp et al. 1993).

Recent observations from the HXT/YOHKOH telescope exhibit similar remote brightenings at X-ray wavelengths (Masuda et al. 1994). Fig.6a shows the time profile of the hard X-ray burst measured in the four HXT energy bands (resp 14-23; 23-33; 33-53 and 53-93 keV). In the highest energy channel, the event consists mainly of two peaks at 2236:20UT and at 2237:40UT, the first peak exhibiting the hardest spectrum. Images obtained in the energy bands 14-23 and 33-53 keV for both peaks as well as for a precursor peak around 2233:00UT are shown in Fig.6b. We shall focus here on the image at higher energy for which the X-ray thermal contamination is expected to be negligible. In the precursor phase, a double structure is observed. The source structure drastically changes at the time of the hardest X-ray peak and a new compact source appears 40" south- west of the first position. This shows another example of the activation of a remote site at the time of the production of the most energetic particles. During the second softer major spike, the emission is mainly produced in the region which was activated during the precursor phase.

As a conclusion, as revealed from spatially resolved observations, the production of non-thermal particles in the solar atmosphere occurs in a complex magnetic environment. In the course of the flare, many different small scale as well as large scale magnetic structures are traced by the energetic electrons. These observations support the concept that a flare may arise from the organized building of successive elementary energy releases in many different magnetic structures. Furthermore, changes in the characteristics of the accelerated particles may be found in connection with modifications of the spatial configuration traced by the particles.

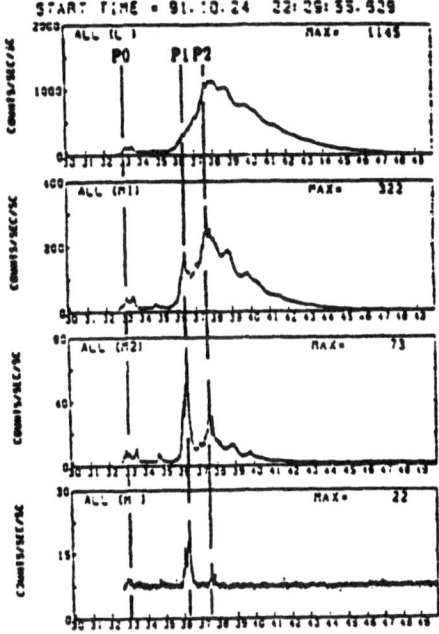

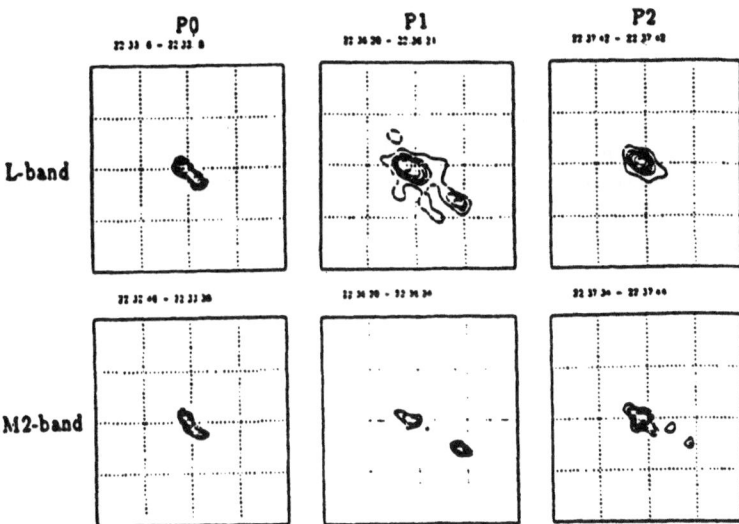

Fig. 6. from Masuda et al. 1994) Time profiles of the 24 October 1991 flare measured in the four YOHKOH/HXT energy bands (a). Hard X-ray images in the L (upper) and M2-bands (lower) at the times of the three peaks of the high energy channel: P0, P1, P2. The field of view is 126" x 126" (b).

4 Distribution in Heights and Spatial Scales of Electron Acceleration Sites

From hard X-ray and radio observations it is found that electron acceleration occurs in a wide range of heights in the solar atmosphere. Studies based on the frequency drifts of decimeter and centimeter type III bursts as well as on the common start frequency of pairs of opposite drifting bursts in the 0.1 - 3.0 GHz range indicate that electron beams arise from a medium with a density ranging from 10^9 to 10^{11} cm^{-3} (e.g. Benz and Aschwanden, 1991, Aschwanden et al. 1995b). This suggests acceleration heights of at least a few 10^4 km. It is generally believed that the bulk of the non-thermal particles produced in flares is also produced at heights close to 10^4km (e.g. for reviews Kane 1987, Lin 1993, Klein 1994). This is consistent with some X-ray images from HXT/YOHKOH suggesting that the emission - and thus the energy release - appears first at the top of magnetic loops (e.g. Takakura et al. 1993). There are however other observations showing that significant acceleration of energetic electrons (a few tens of keV) occurs at heights as high as $2\ 10^5$ km, such as the combined observations of an X- ray flare occulted for ISEE-3 and GOES spacecrafts and non occulted for PVO (Kane et al, 1992).

Magnetic reconfiguration and reconnection is a usual scenario proposed for the production of energetic electrons in the low corona. More recently, combined observations of a long duration soft X-ray flare imaged with the soft X- ray telescope (SXT) aboard YOHKOH and of non thermal radio bursts with the Nançay Radioheliograph have provided the first indication that large scale magnetic reconnection may take place high in the corona on time scales of a few hours (Manoharan et al. 1996). Fig.7a (from Manoharan et al. 1996) displays the SXT images just before, during and at the end of the X-ray flare which is observed by GOES to start at 0948 UT and to last until 1600 UT. Before the flare onset (top-left) a system of twisted magnetic loops is seen at the flare site. This system is observed to expand during the flare (top right). At least two successive twisted expanding loops are followed to a large angular distance from the flare center (middle plots of Fig.7a showing the field of view indicated by a white box in the top frames). While the first loop (Loop 1) starts to expand with a velocity of around 300 kms^{-1} a few minutes after the flare onset (Fig.7b), a second loop system (Loop 2) is observed to expand after 0956 UT. Fig.7b shows the expansion rate of these loops which can be followed to a distance of 8" from the active center before they fade out in X-rays. Shortly after the fading of Loop 2, a series of sporadic radio bursts are observed in close spatial and temporal association with an extrapolated further expansion of this loop (bursts A to F indicated on Fig.7b). During this phase radio bursts indicated by letters a to f on Fig.7b are still observed in close spatial connection with the active region. This clearly shows that non thermal energy release can occur on the same time on small scales related to the active region and on larger scales involving magnetic structures outside the active region, i.e. at widely different heights.

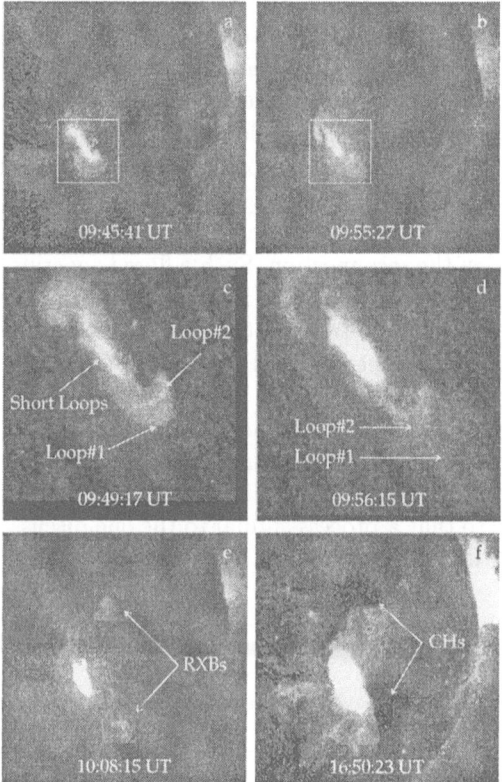

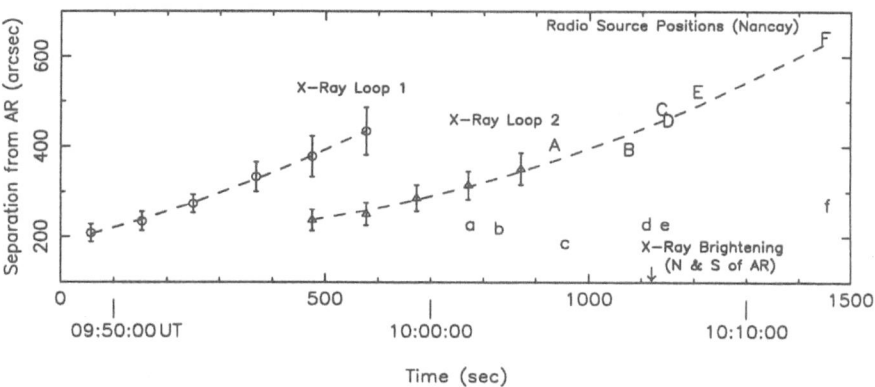

Fig. 7. (from Manoharan et al. 1996) Top Soft X-ray images at 6 different times for the long duration flare of 25 October 1994. Images c and d are partial frame images of the active region corresponding to the field of view indicated by a white box in (a) and (b). Bottom Positions of the leading edges of X-ray loops and location of radio bursts with respect to the center of the active region as a function of time. The vertical bars indicate the typical width of the loop.

Finally, at 1008 UT, two remote X-ray brightenings (RXBs) are observed (Fig.7a, bottom left). The present observations have been interpreted by the authors as an evidence that magnetic reconnection occurs at two levels in the corona: the first one in the active region and the second one involving an interaction between active region loops and surrounding magnetic fields. This is supported by the magnetograph observations which indicate that the remote X-ray brightenings overlie magnetic regions with opposite polarities. This suggests the existence of outer loops between these quiet regions which could be involved in the reconnection. Such a flare triggering of the large scale magnetic reconnection could provide new insight into the understanding of the injection of energetic electrons in the interplanetary space. Furthermore, the interaction found between active region loops and surrounding magnetic fields could provide some support to the suggestion by Kane et al. (1995) that the "giant" flares detected by Ulysses imply that the non thermal energy release occurs in a volume in the corona larger than the one associated with an active region. Part of these "giant" flares are associated with long duration soft X-ray events so that a process similar to the one discussed in this Section could be at work for these flares.

5   Energy Spectra of Flare Accelerated Electrons

5.1   Evidence of Different Accelerated Electron Components?

The exact shape of the accelerated electron spectra is in fact poorly known. Most of the hard X-ray observations have indeed been performed with detectors with a poor spectral resolution. X-ray spectra have been described either as single or double power laws in energy or as resulting from one or many thermal components (see e.g. Vilmer 1987 for a review). To deduce constraints on the non thermal electrons from X-ray observations, the electron spectrum giving rise to X- ray bremsstrahlung is usually characterized by a single power law in energy and the emission is assumed to be produced in a thick target approximation (e.g. Brown 1971). This allows to estimate the energy contained in non thermal electrons above e.g. 25 keV which is from e.g. HXRBS/SMM observations in the range 10^{28} to 10^{32} ergs (Crosby et al. 1993). This content can reach a value of 10^{34} ergs in the case of the "giant" flares observed with ULYSSES (Kane et al. 1995).

There is so far a unique observation of a solar X-ray burst performed with a detector with high spectral resolution (1 keV) (Lin et al. 1981) which allows to determine, without making any assumption, the shape of the instantaneous emitting electron spectrum through direct inversion of the X-ray spectrum (Johns and Lin 1992). Figure 8a shows on the left the X-ray spectrum integrated over the impulsive phase of the 27 June 1980 event. The electron spectrum resulting from the inversion is shown on the right. It exhibits a "superhot" thermal distribution with a temperature of $T \geq 3\ 10^7$ K (Lin et al. 1981). Above 25 keV the electron spectrum is well represented by a double

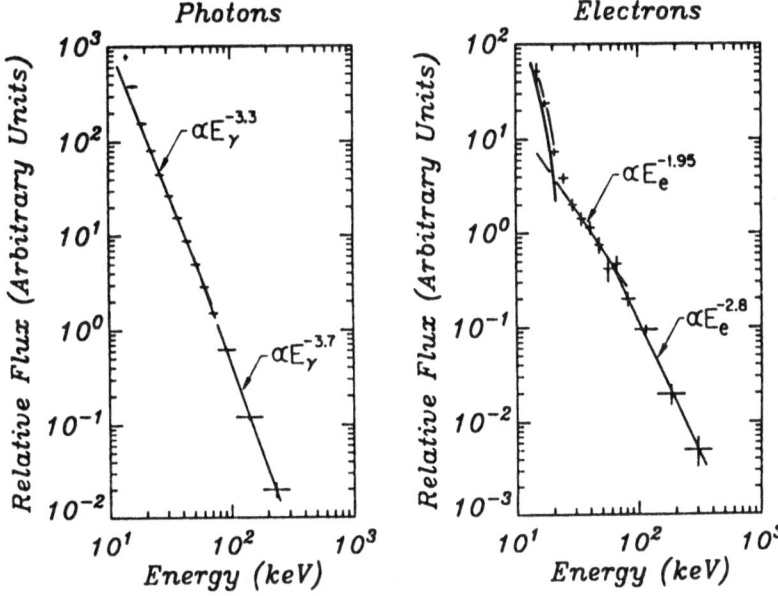

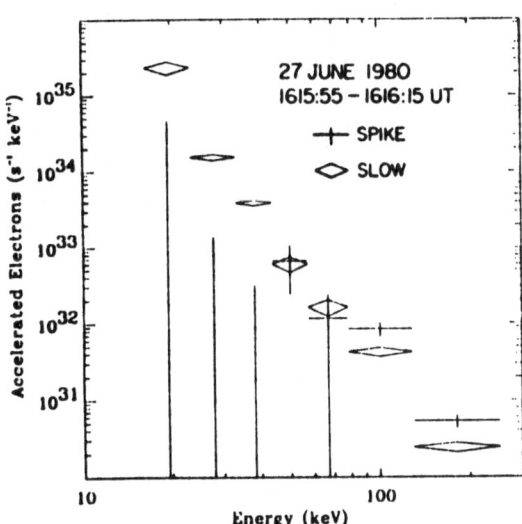

Fig. 8. Left Hard X-ray spectrum observed for the 27 June 1980 event by HIREX between 1615:36 and 1616:50 UT. Right Deconvolved X-ray emitting electron spectrum (from Johns and Lin 1992) (a). Spectra of the accelerated electrons during the spike (crosses) and slowly varying component (diamonds) (from Lin and Johns 1993)

power law with a steepening above 80 keV. Assuming that Coulomb collisions are the dominant energy loss mechanisms in the X-ray emitting source
(i.e. $n \geq 10^{11}$ cm^{-3}), Lin and Johns (1993) derive the accelerated electron
population in time steps of 5 seconds through the flare. They use a continu-
ity equation relating the instantaneous electron number and the electron flux
injected in the emitting source. The X-ray time profile has been interpreted
by Lin and Schwartz (1987) as resulting from a slowly varying component
(time scale of 1 minute) with superposed impulsive spikes (10 s). While the
impulsive spikes are clearly seen in the X-ray flux down to 30 keV, they
are detected in the accelerated electron spectrum $F(E,t)$ only above 78 keV
(Lin and Johns 1993). The X-ray spikes thus appear to result mainly from
high energy electrons which produce significant bremsstrahlung contribution
at lower energies. Although other interpretations are possible, the authors
conclude that both the temporal and spectral characteristics of the accelerated electrons suggest two non thermal electron components: a slowly varying
one and a spiky one. The accelerated electron spectrum associated with the
spike is then deduced (Fig.8b) assuming that the slowly varying component
keeps a constant spectrum during most of the burst. The accelerated electron
spectrum for the slowly varying component is well represented by a power
law from 20 keV to 200 keV and the accelerated spectrum during the spike
exhibits a peak at 50 keV, has a low energy cut-off below 40 keV and has
a harder spectrum. The rate of energy input in electrons $\geq$ 23 keV is about
6 times smaller for the spike than for the slowly varying component. This
shows that most of the flare energy is contained in low energy electrons independently of the production of higher energy spikes. The peaked electron
acceleration spectrum of the spike is similar to electron spectra observed in
aurorae (see e.g. Mozer et al, 1980 for a review) and could indicate some
high energy electron acceleration by quasi-parallel electric fields . This acceleration mechanism would however not produce the bulk of the electrons. An
alternative interpretation of the X-ray time profile suggested by Sect. 2 and 3
would be that rather than resulting from two different electron populations,
the burst is a succession of different elementary peaks. Without simultaneous spatial resolution at X-ray wavelengths, it is not possible to decide which
interpretation is the most plausible one.

5.2   Bremsstrahlung and Synchrotron Emitting Electrons: Energy Spectra at Low and High Energies

Spectral hardening of the photon spectra (and thus probably of the emitting
electron spectra) has been reported in a few events above 300 keV in the
SMM (see e.g. Dennis 1988, Marschhaüser et al. 1994) and HINOTORI observations (Yoshimori 1989). A few events observed by PHEBUS/GRANAT
with a photon spectrum extending up to 10 or 100 MeV also exhibit a clear
hardening of the spectrum at high energies: e.g. the 01 June 1991 event (Barat
et al. 1994) and the 11 June 1990 event (Fig.9b) (Trottet et al. 1996). The

four lower plots of Fig.9a display the time evolution of the hard X-ray/γ-ray count-rates detected in 4 energy bands for the 11 June 1990 event. Despite of the short duration of the burst, it consists of a succession of peaks of typical duration of the order of 10 seconds marked a to f on the 0.12-0.22 MeV count-rate time profile. Among the four most intense peaks (b, c, d, e) only peak d shows a significant emission up to 56 MeV while peak c is associated with a marginally significant excess above 10 MeV around 0943:22 UT. Spectral analysis performed for count spectra accumulated on time intervals corresponding to these five peaks has been performed. The spectral fits using the detector response function lead to the following results:

1. During peaks c, d and e, the observed spectra are consistent with a double power law photon spectrum. This is illustrated in Fig.9b (from Trottet et al, 1996) for peaks c and d. The solid curves drawn through the data points represent the best fit photon spectrum models while the observed count-spectra have been converted to photon spectra (plotted data points). During peaks c and e, the photon spectrum keeps a similar shape with a break energy E_b around 0.35 MeV, a slope at lower energies γ_1 $\simeq$ 4.3±0.2 and at higher energies $\gamma_2 \simeq$ 2.0±0.1. During peak d, the observed spectrum is harder. However, the main differences are observed in the lower part of the spectrum since a hardening of the low energy part is observed ($\gamma1 \simeq$ 2.9± 0.1) as well as a shift of the break energy towards high energies (E_b = 0.7 ± 0.1 MeV). The photon spectrum at high energies keeps a similar slope ($\gamma2 \simeq$ 1.7 ± 0.1) (Trottet et al. 1996).

2. The count-rates during the other peaks are too low at high energies to show up any hardening in the photon spectra. The low energy part of the spectrum is always found to be steeper than 4 (Trottet et al. 1996).

This hard X-ray/γ-ray event is associated with a microwave/millimeter wave burst observed with the Bern polarimeters in the whole band from 3 to 50 GHz. The two upper curves of Fig.9a shows the time profiles observed at 35 GHz (optically thin part of the spectrum) and at 11.8 GHz (around the turnover frequency). At 50 GHz, the time profile is similar to that observed at 35 GHz, apart from peak a which is barely detectable. For most of the flare (from ~ 0943 UT to 0944 UT), the microwave emission is characterized by a turnover frequency between 11.8 and 19.6 GHz. The microwave spectrum is characterized by a radio spectral index between 35 and 50 GHz around - 1.5 during peaks b, c, d and during the first half of peak e (see e.g. Fig.9c for peaks c and d). A simple attempt to relate the slope of the bremsstrahlung emitting electrons deduced from hard X-ray/γ-ray observations to the slope of the microwave/millimeter emitting ones in the optically thin regime has been performed following the suggestions of Kai et al. (1985), Kai (1986), Kosugi et al. (1988) and Ramaty et al. (1994). It indicates that the electrons producing the high frequency emission (e.g. above 19 GHz)cannot be related to the low energy part of the bremsstrahlung emitting electrons which has a too soft energy spectrum. On the other hand, during most of the event,

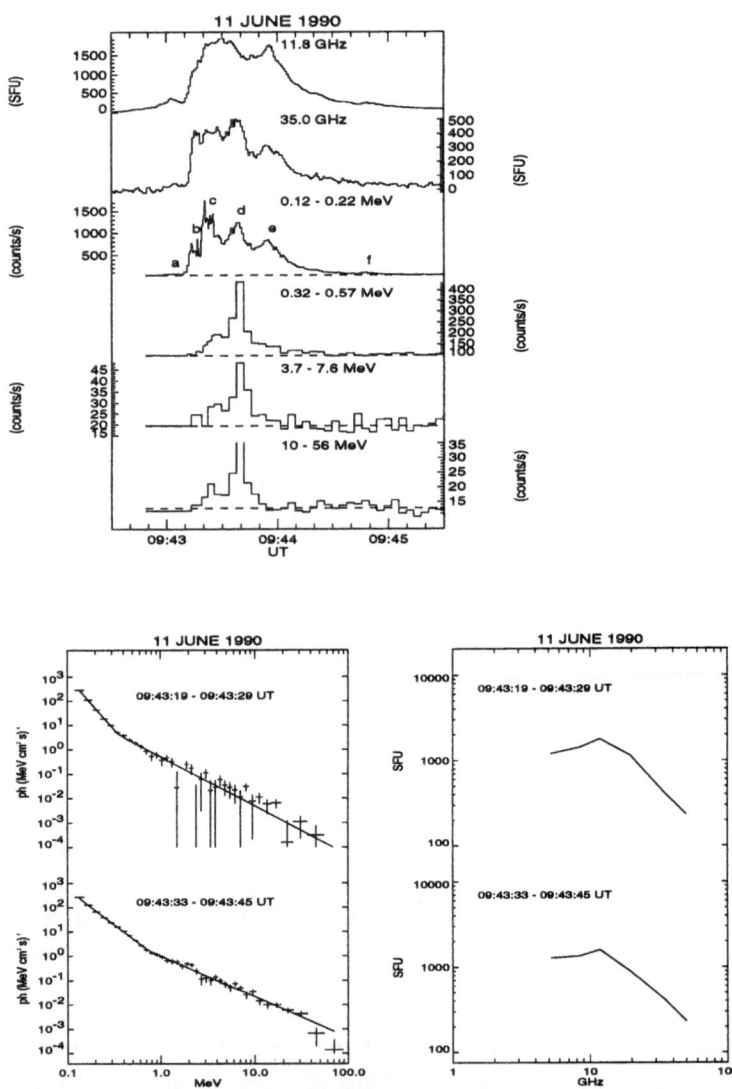

Fig. 9. (adapted from Trottet et al. 1996) Time profiles of the 11 June 1990 event measured in four energy bands with PHEBUS/GRANAT and at two microwave frequencies with the Bern polarimeters (a). Hard X-ray/γ-ray photon spectra from PHEBUS/GRANAT (Barat, private communication) during peaks c and d. The solid lines represent the best fit double power law photon spectrum(b). Microwave spectra from the Bern polarimeters (Magun, private communication) during peaks c and d (c).

the spectral index of the high frequency emitting electrons is related to the one of the high energy bremsstrahlung emitting ones. Even during peak a, the value of the radio spectral index measured between the two optically thin frequencies of 19.6 and 35 GHz (which is ~ -1.5) indicates that the high energy hard electron spectrum is already present in the radio emitting source, even if it is not yet detected by its bremsstrahlung emission due to too low photon count-rates (Trottet et al. 1996).

The above results are in agreement with the suggestions by White and Kundu (1992) and Kundu et al. (1994) that millimeter wave emission (e.g. at 86 GHz) is produced by high energy electrons around 1 MeV character-ized by a very flat spectrum, usually much flatter that the one deduced from X-ray observations below 100 keV (Kundu et al. 1994). This millimeter emis-sion originating from high energy electrons is observed in flares of all sizes, sometimes even in the beginning phase (Kundu et al. 1994). Such observa-tions led Kundu et al. (1994) to assume that millimeter wave observations are the most sensitive diagnostics of MeV electrons and that these high energy electrons could constitute a different population from the one of the elec-trons around a few hundred keV and be produced by a separate acceleration mechanism. The observations reported here from PHEBUS/GRANAT do not support nor contradict by themselves the hypothesis of several accelerated electron populations observed at low and high energies. They bring some ob-servational support to the fact that high frequency gyrosynchrotron emissions can sometimes be related to the highest energy part of the bremsstrahlung emitting electrons. The coordinated analysis of the PHEBUS/GRANAT and of the Bern polarimeters observations represents the first attempt to relate bremsstrahlung and gyrosynchrotron emitting electrons using observations in a wide energy range going from 100 keV to 100 MeV as well as a wide frequency range. Such an approach is in fact necessary to provide an unam-biguous analysis of the electron spectra at the origin of both X-ray, γ-ray and cm/mm wave emissions.

6   Confrontation of the Experimental Data with the Predictions of the Avalanche Models of Flares

As discussed in Sect.2.1, the occurrence distribution of the peak count-rates of HXR pulses above 50 keV can be represented by a power law with a slope similar to the one obtained for the frequency distribution of global solar X-ray flare peak count-rates from different experiments (e.g..Crosby et al. 1993, Lu et al. 1993, Bromund et al.1995, Biesecker et al. 1994). Such power law distri-butions hold for several decades for different global parameters of hard X-ray events (e.g. peak count-rate, duration, total energy contained in non-thermal electrons above 20 or 25 keV) and have been interpreted in the context of the evolution of a complex system towards a self organized state (see e.g. Lu and Petrosian 1989, Lu and Hamilton 1991, Lu et al. 1993; Lu 1995a 1995b,

Vlahos et al. 1995, Georgoulis and Vlahos 1996). In this approach, solar active regions are modeled as driven dissipative systems in which the long term statistical behavior is characterized by a power law distribution of energy dissipating avalanches. Solar flares can thus be understood as avalanches of many small reconnection events. Different numerical simulations have been performed and have predicted power law distributions for the different observable parameters with slopes consistent with the observations (see e.g. Trottet and Vilmer 1996 for a review). Furthermore, the comparison of the results of a numerical simulation and of the observed frequency distributions from ISEE3/ICE led to the prediction that power law occurrence distributions should continue downward to smaller scales with the same slope, down to a typical energy of $3\ 10^{25}$ ergs and duration of 0.3s (Lu et al. 1993). These lower limits are obtained under the crude assumption that each individual reconnecting cell is identical. They represent in this approach the characteristic energy and time scale of an elementary avalanche which, according to the prediction, should occur on a typical length scale of 400 km. The observations reported in Sect.2.1 of a quasi-systematic detection of time structures of duration of a few hundred milliseconds in solar hard X-ray bursts and of a scale invariance between the occurrence distributions of the peak count-rates for HXR pulses and hard X-ray bursts thus provide a strong support to the avalanche models of flares. The time structures of a few hundred milliseconds (also reported by Kaufmann (1985)) would represent the elementary bursts building a flare but would already result from the clustering of many individual fragments. Furthermore, the observations that there is no unique value of characteristic time scales in a flare or from one flare to the other provide another support to this type of models.

The prediction of the "avalanche" model developed by Lu et al. (1993) also suggests that the typical size of an avalanche lasting a few hundred milliseconds would be of the order of 400 km (~ 0.5"). With these values as guidelines, it must be expected that movements of the emitting sources of non-thermal radiation would not be observed on time scales of a few hundred milliseconds, given the spatial resolution of the observations. However, if a flare is built from many avalanches in a complex system, spatially resolved observations of non-thermal radiation should reveal variations in space on time scales larger than a few hundred milliseconds (namely on time scales of a few seconds to a few tens of seconds). Such observations have been reviewed in Sect.3. In particular, millimeter radio observations obtained at 48 GHz and reported by Correia et al. (1995) suggest that an impulsive solar burst is a superposition of small events spread both in time and space. However, subsecond fluctuations observed in the time profiles cannot be spatially resolved and are produced in a small area of diameter of about 3" (see Sect.3.1). If the fast pulses are elementary avalanches of energy release, the spatial size of such avalanches is then, as expected from the prediction of Lu et al. (1993), less than 3".

7   Conclusion and Prospective

Hard X-ray and radio observations from radiation of non thermal electrons, have been reviewed in this paper. From the observations obtained at high temporal resolution and/or with spatial resolution, there is some indication that the non thermal energy release is fragmented in both space and time. Time structures of a few hundred milliseconds are quasi systematically detected in solar hard X-ray bursts and spatially resolved observations suggest that flares result from the superposition of successive short-lived events emitted at different sites in spatially distinct small scale as well as large scale magnetic structures. Furthermore, studies of the distribution of fine time structures in hard X-ray bursts show scale invariance with similar distributions of global parameters of hard X-ray flares. The confrontation of the prediction of the numerical simulations of the "avalanche" models with the observed distributions of global parameters of hard X-ray flares (see Sect.6) together with the ensemble of observations at high temporal and/or spatial resolution reviewed here strongly support that a flare consists of an "avalanche" of many fragments of energy release, such as small reconnection events. However, observational work performed along these guiding lines must be pursued with the available past and new data sets to unambiguously establish such an idea, since individual reconnection events are far from being observed. In particular, the examples presented in Sect.2.1 of spatially resolved observations suggesting that flares result from successive short-duration events emitted at different sites indicate that the separation of the emitting components is small, of the same order as the spatial resolution of the radio interferometers, and that the changes of the position of the predominant components occur on short time scales (a few seconds). This may explain why such rapid changes in the morphology of the emitting sources of non-thermal radiation are scarcely reported in the literature. To search for them in more events, it would be necessary to re-examine already processed images and search for rapid displacements (on time scales of a few seconds) of the barycenter of the image. This is a potential study which could be performed in the centimeter/millimeter domains with the Nobeyama Radioheliograph observations at 17 GHz (e.g. Enome, 1995), by examining the temporal evolution of the phases recorded with different baselines by BIMA at 86 GHz (e.g. White and Kundu, 1992) and pursued with the ITAPETINGA observations. Finally, at X-ray wavelengths, such rapid changes of the centroid of the emitting source from one elementary burst to the other could be searched for, using the observations of the HXT/YOHKOH telescope with the basic time resolution of 0.5s (Kosugi et al, 1991). On the theoretical point of view, some effort must be devoted in a further development of these "avalanche" models of flares which should predict how the energy is released among the different kinds of particles (including ions which have not been discussed in this review) and how the energy release distribution can evolve in time and space. In particular, one must explain the connection sometimes observed between the characteristics of the energetic particles and the modifications of the different small scale as well as large scale magnetic structures traced by the emitting particles. Other developments and tests of these "avalanche" models can be found in Bastian and Vlahos (1996).

Radio and X-ray observations also show that the production of non thermal electrons occurs in a wide range of heights. Not only the analysis of small scale phenomena are thus necessary to understand this production. Magnetic reconnection can indeed occur simultaneously at two levels in the corona: in the active region and at greater heights with the interaction between expanding loops of the active region and surrounding magnetic fields. In the near future, the observations performed with the new version of the Nançay Radioheliograph will allow to make real time 2D images of the solar corona at five heights and thus to probe large scale as well as medium scale structures traced by energetic electrons. Together with observations providing information on the non thermal particles in smaller scale structures, these observations will be crucial to get further constraints on the role of the magnetic structures at different scales in the production of non thermal electrons and on the link between the energetic electron populations observed at different coronal heights. Such observational constraints should also be considered in the future developements of models of energy release.

References

Akimov V.V., GAMMA1 Team (1991): in Proc 22d Internat Cosmic Ray Conf (Dublin) **SH2.3-6**, 73

Aschwanden M.J., Benz A.O., Dennis B.R. et al. (1995b): ApJ **455**, 347

Aschwanden M.J., Schwartz R.A., Alt D.M. (1995a): ApJ **447**, 923

Barat C., Trottet G., Vilmer N. et al. (1994): ApJ **425**, L109

Bastian T.S., Vlahos L. (1996): this volume

Benz A.O., Aschwanden M.J. (1991): in Eruptive Solar Flares, ed. by Z. Svestka, B.V. Jackson, M.E. Machado, Lect. Notes in Physics **399**, 361

Biesecker D.A., Ryan J.M., Fishman G.J. (1994): in High Energy Solar Phenomena - a new Era of Spacecraft Measurements, ed by J.M. Ryan, W.T. Vestrand, AIP Conf. Proc. **294**, 183

Bromund K.R., McTiernan J.M., Kane S.R. (1995): ApJ **455**, 733

Brown J.C. (1971): Solar Phys. **18**, 489

Chupp E.L. (1996): in High Energy Solar Physics, ed by R. Ramaty, N. Mandzhavidze, X.M. Hua, AIP Conf. Proc. **374**, 3

Chupp E.L., Trottet G., Marschhaüser H. et al. (1993): Astron. Astrophys.**275**, 602

Correia E., Costa J.E.R., Kaufmann P. et al. (1995): Solar Phys. **159**, 143

Crosby N.B., Aschwanden M.J., Dennis B.R (1993): Solar Phys **143**, 275

de Jager C., de Jonge G. (1978): Solar Phys. **58**, 127

Dennis B.R. (1988): Solar Phys. **118**, 49

Enome S. (1995): in Coronal Magnetic Energy Releases, ed by A.O. Benz, A. Krüger, Lect. Notes in Physics **444**, 35

Georgoulis M.K., Vlahos L. (1996): ApJ Letters **469**, L135

Herrmann R., Magun A., Costa J.E.R. et al. (1994): Astron. Astrophys., in press

Johns C.M., Lin R.P. (1992): Solar Phys. **137**, 121

Kai K. (1986): Solar Phys. **104**, 235

Kai K., Kosugi T., Nitta N.(1985): Publ. Astron. Soc. Japan **37**, 155

Kanbach G., Bertsch D.L., Fichtel C.E. et al. (1993): Astron. Astrophys. Suppl. Ser. **97**, 349

Kane S.R. (1987): Solar Phys **113**, 145

Kane S.R., Hurley K., McTiernan J.M. et al. (1995): ApJ **446**, L47

Kane S.R., McTiernan J.M., Loran J. et al.(1992): ApJ **390**, 687

Kaufmann P. (1985): Solar Phys. **102**, 97

Kiplinger A.L., Dennis B.R., Emslie A.G. et al. (1983): ApJ **299**, 285

Klein K.L. (1994): in Advances in Solar Physics, ed. by G. Belvedere, M. Rodono, G.M. Simnett, Lect. Notes in Physics **432**, 261

Klein K.L., Anderson K.A., Pick M. et al. (1993): Solar Phys. **84**, 295

Kosugi T., Dennis B.R., Kai K. (1988): ApJ **324**, 1118

Kosugi T., Makishima T., Murakami T. et al. (1991): Solar Phys. **136**, 17

Krucker S., Aschwanden M.J., Bastian T.S. et al. (1995): Astron. Astrophys. **302**, 551

Kundu M.R., White S.M., Gopalswamy N. et al. (1994): ApJ Supp. Series **90**, 599

Leikov N.G., Akimov V.V, Volzhenskaya V.A. et al. (1993): Astron. Astrophys. Suppl. Ser. **97**, 345

Lin R.P. (1993): in Fundamental Problems in Solar Activity, ed by M. Pick, M.E. Machado, Adv. Space Res. **13-9**, 265

Lin R.P., Johns C.M. (1993): ApJ **417**, L53

Lin R.P., Schwartz R.A. (1987): ApJ **312**, 462

Lin R.P., Schwartz R.A., Pelling R.M. et al. (1981): ApJ **251**, L109

Lu E.T, Hamilton R.J, McTiernan J.M. et al. (1993): ApJ **412**, 841

Lu E.T. (1995a): ApJ **446**, L109

Lu E.T. (1995b): ApJ **447**, L416

Lu E.T., Hamilton R.J. (1991): ApJ **380**, L89

Lu E.T., Petrosian V. (1989): ApJ **338**, 1122

Manoharan P.K., van Driel-Gesztelyi L., Pick M. et al. (1996): ApJ **468**, L73

Marschhaüser H., Rieger E., Kanbach G. (1994): in High Energy Solar Phenomena - a new Era of Spacecraft Measurements, ed by J.M. Ryan, W.T. Vestrand, AIP Conf. Proc. **294**, 171

Masuda S., Kosugi T., Sakao T. (1994): in X-ray Solar Physics from YOHKOH, ed. by Y. Uchida et al., 123

Mozer F.S., Cattell C.A., Hudson M.K. et al. (1980): Space Sci. Rev. **27**, 155

Pelaez F., Mandrou P., Niel M. et al. (1992): Solar Phys **140**, 121

Ramaty R., Schwartz R.A., Enome S. et al. (1994): ApJ **436**, 941

Raoult A., Pick M., Dennis B.R. et al. (1985): ApJ **299**, 1027

Rieger E. (1994): ApJ Supp. Series **90**, 645

Rieger E., Marschhaüser H. (1990): in Proc 3d MAX'91/SMM Workshop on Solar Flares : Observations and Theory, ed. R.M. Winglee, A.L. Kiplinger, 68

Takakura T., Inda M., Makishima K. et al. (1993): Publ. Astron. Soc. Japan **45**, 737

Trottet G. (1994): in Fragmented Energy Release in Sun and Stars, ed. by G.H.J. Van den Oord, Space Science Reviews **68**, 149

Trottet G. (1986): Solar Phys. **104**,145

Trottet G., Chupp E.L, Marschhaüser, H. et al. (1994): Astron. Astrophys.**288**, 647

Trottet G., Vilmer N. (1996): in Solar and Heliospheric Plasma Physics, Lect. Notes in Physics, to be published

Trottet G., Vilmer N., Barat C. et al. (1993): Adv. Space Res **13-9**, 171

Trottet G., Vilmer N., Barat C. et al. (1996): to be submitted

Van Beek H.F., de Feiter L.D., de Jager C. (1974): Space Res. **14**, 447

Vestrand W.T., Forrest D.J., Chupp E.L. et al. (1991): in Proc. 22d Internat. Cosmic-Ray Conf. (Dublin), **SH2.3-5**, 69

Vilmer N. (1987): Solar Phys. **111**, 207

Vilmer N., Trottet G., Barat C.etal. (1994): in Advances in Solar Physics, ed. by G. Belvedere, M. Rodono, G.M. Simnett, Lect. Notes in Physics **432**, 197

Vilmer N., Trottet G., Verhagen H. et al. (1996): in High Energy Solar Physics, ed by R. Ramaty, N. Mandzhavidze, X.M. Hua, AIP Conf. Proc. **374**, 311

Vlahos L. (1994): in Fragmented Energy Release in Sun and Stars, ed. by G.H.J. Van den Oord, Space Science Reviews **68**, 39

Vlahos L., Georgoulis M., Kluiving R. et al. (1995): Astron. Astrophys. **299**, 897

White S.M., Kundu M.R. (1992): Solar Phys. **141**, 347

Wülzer J.P., Canfield R.C., Rieger E. (1990): in Proc 3d MAX'91/SMM Workshop on Solar Flares : Observations and Theory, ed. R.M. Winglee, A.L. Kiplinger, 149

Yoshimori M. (1989): Space Sci. Rev. **51**, 85

Radio Observations of the Quiet Sun and Their Implications on Coronal Heating*

Costas E. Alissandrakis[1] and Giorgio Einaudi[2]

[1] Section of Astro-geophysics, Department of Physics, University of Ioannina, GR-45110 Ioannina, Greece
[2] Physics Department, University of Pisa, Piazza Torricelli 2, 56126 Pisa, Italy

Abstract. We discuss the actual and potential contribution of radio techniques in the study of the structure and the dynamics of the corona as well as in the problem of coronal heating. Radio observations provide powerful diagnostics of the physical conditions in the transition region and the corona. Recent observational and theoretical results are presented and their implications are discussed. The prospects of further observations, in conjunction with other wavelength ranges are given.

Résumé. Les observations radio constituent un puissant diagnostic de la région de transition et de la couronne solaires. Nous montrons comment de telles observations, présentes ou futures, contribuent de façon significative à l'étude tant de la structure et de la dynamique que du chauffage de la Couronne. Dans ce contexte, nous tirons les conséquences de travaux expérimentaux et théoriques récents. Nous soulignons, dans un but prospectif, l'intérêt de combiner de nouvelles observations radio avec des observations obtenues dans d'autres domaines de longueur d'onde.

1 Introduction

The solar atmosphere is traditionally divided into active and quiet regions. This distinction is more clear at the photospheric level, where active regions stand out as quite distinct structures against quiet regions, exhibiting significant concentrations of magnetic flux with spatial scales of the order of 10^5 km and temporal scales of the order of a month. However, already at the chromospheric level, it becomes obvious that the quiet sun is not that quiet after all. The chromospheric network, with spatial scales of about 3×10^4 km and temporal scales of about 10 min is already a dynamic system with network structures such as mottles and spicules exhibiting a complex behaviour, involving the release of significant amounts of energy. Higher up, in the corona, the dynamic characteristics of the quiet sun are reinforced, with bright points behaving like tiny active regions and with phenomena such as the large scale restructuring of the magnetic field which sometimes lead to Coronal Mass Ejections.

As we learn more both about active regions and the quiet sun we realize that, in spite of the difference in scales, quiet and active regions have many

---

* Working Group Report

things in common as far as physical processes are concerned. Thus, a lot of what we have learned from the study of active regions can be applied to the quiet sun and vice versa. This does not imply that the quiet sun does not exist as an independent entity, governed by magnetic fields at all scales, from below the resolution limit up to more than one solar radius. By contrast, active regions are characterized by high flux, intermediate scale magnetic fields with a predominance of the bipolar component.

The purpose of this working group was to assess the actual and potential contribution of radio techniques in the study of the structure and the dynamics of the corona as well as in the classic problem of coronal heating. Of course, if one adopts Scudder's (1992) view according to which the corona is the consequence of velocity filtration from a suprathermal tail at the base of the transition region where the plasma is nearly fully ionized and optically thin, there is no need for a specific heating mechanism; however, here we will follow the more traditional approach, which associates coronal heating with the process of energy transport from the lower layers and its subsequent dissipation.

The radio domain is well known for its advantages, such as the possibility to observe the corona from the ground, the capability to probe the solar atmosphere from the chromosphere up to the middle corona, the simplicity of (most) physical processes involved in the radio emission and the sensitivity in the magnetic field. Equally well known are its limitations, for example in high spatial resolution (which, however, is no longer a major problem for short cm-λ) and in the influence of refraction effects and scattering. Recent reviews on the subject have been given by Alissandrakis (1994a; 1994b). Although our working group definitely did not solve the problems, we have the feeling that we did make some progress in putting together the observational data and pointing out some outstanding questions.

2   Radio Diagnostics of the Corona

2.1   One-Dimensional Models

Radio observations give the brightness temperature as a function of wavelength and position on the solar disk. On the basis of this information, the electron temperature and density as a function of height can be derived.

Extensive work on the subject has been carried out by the Pulkovo group, presented here by G. Gelfreikh, using the RATAN-600 radio telescope in the wavelength range of 2-32 cm (Borovik et al., 1989; 1992; 1993); the instrumental resolution was sufficient to discriminate between active regions, the quiet sun and coronal holes. V. M. Bogod and A. S. Grebinskij pointed out to the working group the effects of refraction, at longer cm wavelengths in particular (Bogod and Grebinskij, 1996). They also presented computations of the Differential Emission Measure (Fig. 1) for quiet regions and coronal holes (Grebinskij, 1987), which showed a considerable deficit of matter in the low transition region (around $\log \tau = 4.5$); their model gives a DEM almost two orders of magnitude below that determined from EUV observations by Raymond and Doyle (1981). It is interesting to note here that Zirin et al. (1991) modeled the Owens Valley quiet sun data (1.7 to 21 cm) with an optically thick chromosphere and an optically thin corona, without any contribution from the transition region.

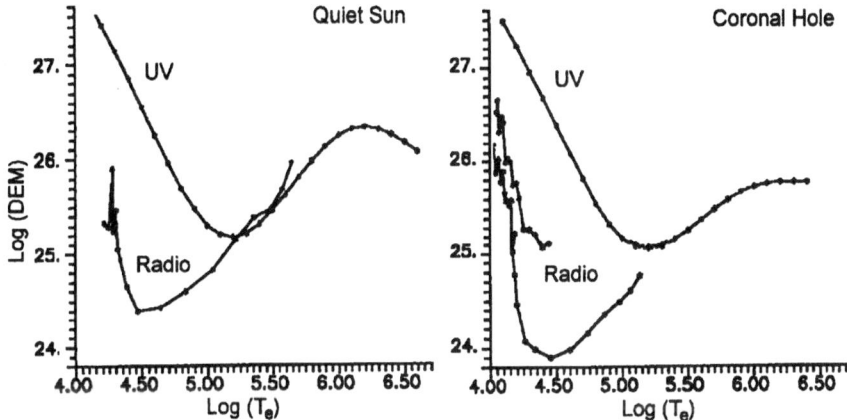

Fig. 1. Differential Emission Measure from radio and UV data for the quiet sun and for coronal holes (courtesy of Bogod and Grebinskij)

2.2   Fine Structure

The problem with 1-D models is that they do not take into account the inhomogeneities, which are known to exist since the early days from their effect on limb brightening. The first high resolution 2-dimensional maps, obtained by Kundu et al. (1979) with the Westerbork Synthesis Radio Telescope at 6 cm, showed a close correspondence of the radio structure with the chromospheric network, confirming earlier suggestions from interferometric observations or 1-dimensional scans (Kundu and Velusamy, 1974, Kundu and Alissandrakis, 1975; Bogod and Korolkov, 1975). More recently, observations at 6 and 20 cm were made by Gary and Zirin (1988) and at 3.6 cm by Gary et al. (1990) using the VLA. The brightness temperature of the network structures increases with wavelength from $10\text{-}20 \times 10^3$ K at 3.6 cm to above 10^5 K at 20 cm, showing that the emission comes from the extension of the chromospheric network into the transition region; at the same time the structures become more diffuse at longer wavelengths (Gary and Zirin, 1988), which apparently reflects the fact that the network fades out in the low corona.

Another emission component has been observed, first at 20 cm (Habbal et al., 1986; Nitta and Kundu, 1988), associated with small bipolar structures and He I dark points; this component has been identified as the radio counterpart of X-ray bright points. These structures are less visible at shorter wavelengths, but they apparently exist down to 3.6 cm; indeed Gary et al. (1990) noticed that bipolar network structures were brighter than average and even observed a bright loop. Finally, microspicules have been reported at 6 cm (Habbal and Gonzalez, 1991).

Two recent sets of VLA observations were presented, by G. Dulk (see Bastian et al., 1996) and by A. Benz (Benz et al., 1996). The observations presented by A. Benz were obtained at 1.3, 2 and 3.6 cm (Fig. 2), together with two deep soft X-ray exposures with the SXT aboard Yohkoh, which showed features 100 times fainter than X-ray bright points. They found a good correlation of X-ray structures with bipolar regions in the photospheric magnetic field, as well as with radio structures. Brightness variations increase, while the cross-correlation coefficient with the absolute value of the magnetic field decreases with wavelength, i.e. with increasing height. According to their model calculations a density increase with rms value of 20% in the chromosphere and 60% in the corona, under constant temperature, can explain the observed standard deviations in intensity.

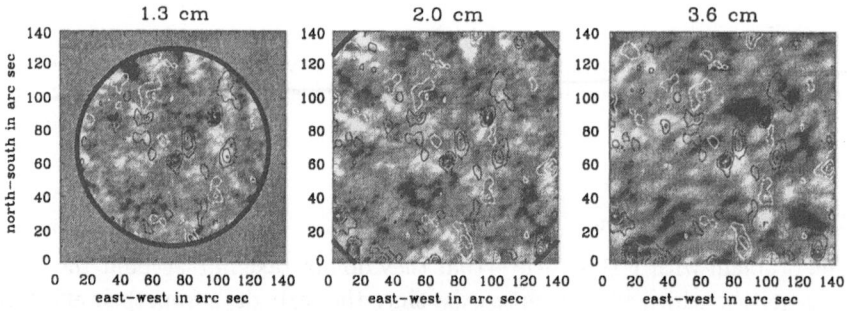

Fig. 2. VLA images of the quiet sun with overlaid contours of a magnetogram from KPNO (from Benz et al., 1996)

G. Dulk showed VLA observations at 1.3 and 2 cm and compared them with magnetograms and He I images. They made accurate measurements of the brightness temperature which is in the range of about $\pm 1000\,\mathrm{K}$ of the mean value. Good spatial correlation was found between the bright radio network and relatively strong magnetic fields. It was pointed out that the brightness temperatures measured with the VLA as well as the ones of Zirin et al. (1991) are significantly lower than those predicted by the classic model of Vernazza, Avrett and Loeser (1981); however a recent model by Avrett (1995), with lower chromospheric temperatures at greater heights than previous models, is consistent both with the microwave and the CO data, but not with the UV/EUV data (Fig. 3). Apparently the UV is biased towards bright structures, while the radio is more sensitive to dark, cooler matter.

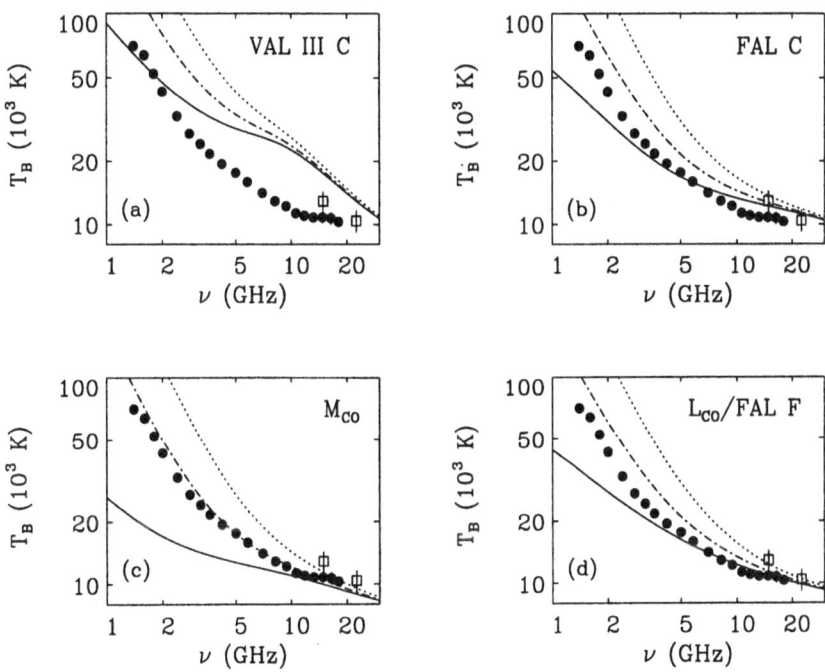

Fig. 3. Microwave spectrum measured by Zirin et al. (1991; filled circles) and with the VLA (open squares). The solid lines are spectra computed from the chromospheric models of Vernazza, Avrett and Loeser (1981; top left), Fontenla, Avrett and Loeser (1993; top right), the model M_{CO} of Avrett (1995, bottom left) and the two component model of Avrett (1995, bottom right). The contribution of the solar minimum corona is included in the dot-dashed curves and that of the maximum corona in the dotted curves. (From Bastian et al., 1996)

2.3 Long Wavelength Imaging Observations

There are practically no imaging observations in the range of 21-70 cm. At longer wavelengths the Nançay Radioheliograph, as well as the now extinct Calgoora and Clark Lake Radioheliographs have provided a wealth of data; more is expected to come from the Giant Meterwave Radio Telescope in India.

Although the spatial resolution at long dm-λ, m-λ and Dm-λ, is of the order of one arc minute or worse, important information has been obtained about the structure of the low and middle corona. Around 160 MHz there is

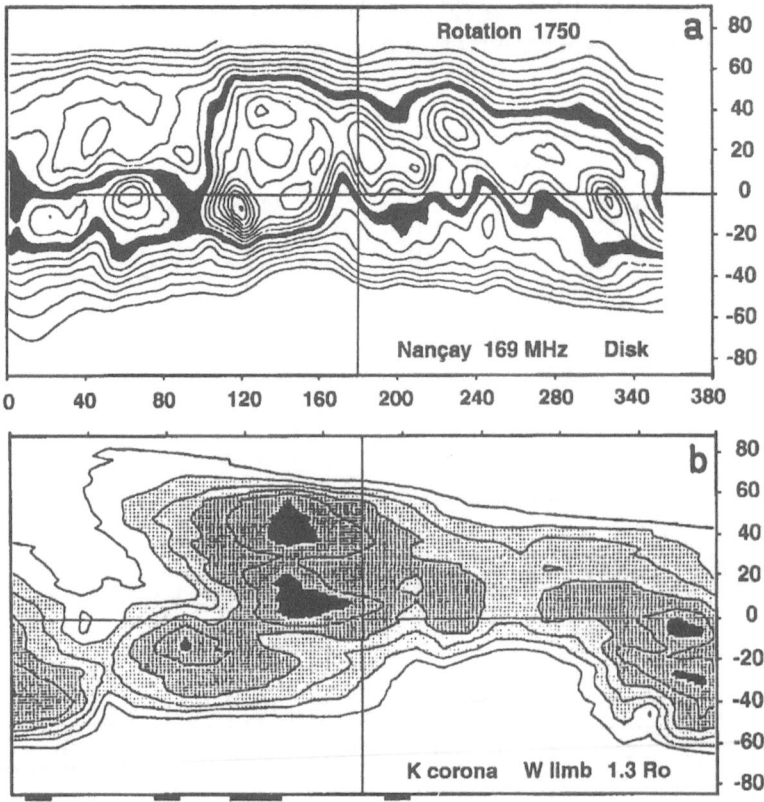

Fig. 4. Nançay synoptic map 169 MHz (top) and in the K-corona at $1.3\,R_\odot$. The border of the coronal plateau is shown in black on the radio map (from Lantos and Alissandrakis, 1996)

practically no trace of emission from active regions, apparently because the density of the associated loops is high enough to place them below the plasma level. Still the emission is often dominated by weak noise storm sources (non-thermal), which occur near active regions and have a brightness temperature as low as 1.5×10^6 K (Alissandrakis et al., 1985). Thermal sources are weaker, 0.5 to 2×10^5 K above the quiet sun background of 9×10^5 K. These sources are located at ≤ 0.2 $R_\odot$ above the photosphere, near the neutral line of the photospheric magnetic field, and have been interpreted as coronal arch systems (see also Alissandrakis and Lantos, 1996). At longer wavelengths (higher up in the corona) streamers become quite obvious, whereas coronal holes have higher contrast at shorter wavelengths and sometimes appear as bright structures at decametric wavelengths.

Lantos et al. (1992) found that all quiet sun sources that they observed at 164 and 408 MHz were located inside the zone of enhanced K-corona

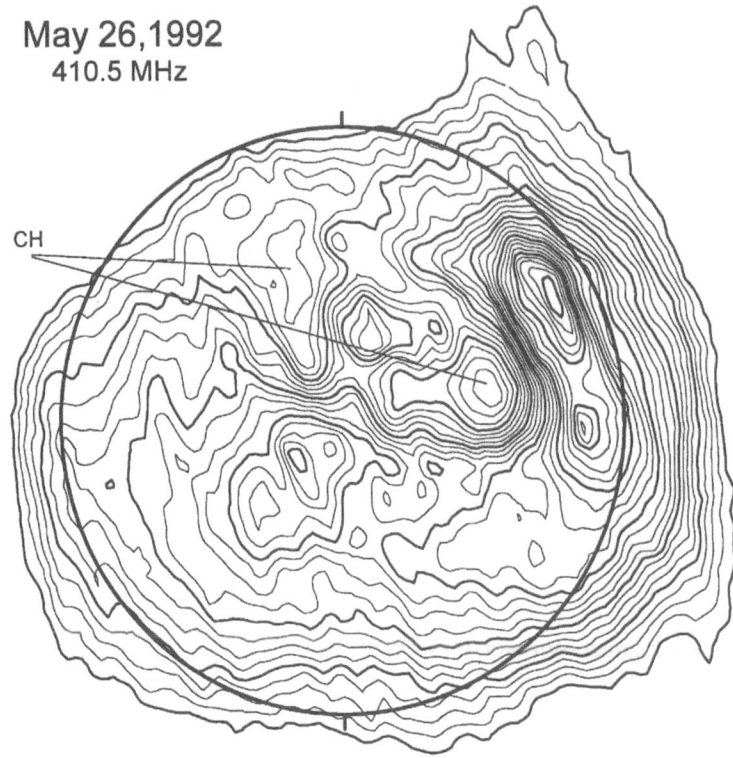

Fig. 5. Nançay map of the sun at 410.5 MHz, for May 26, 1992. Two coronal holes are marked (courtesy of A. Coulais and P. Lantos)

emission. This strong association shows that the quiet solar emission at long wavelengths is associated with the large scale structure of the coronal magnetic field. They also detected a region of enhanced brightness around the local sources which they called *coronal plateau*. The plateau was found to be stable during at least one solar rotation at 164, 327 and 408 MHz and it followed closely the shape of the coronal neutral sheet as observed with the K-coronameter (Fig. 4; see also Lantos and Alissandrakis 1996).

The Nançay Radioheliograph has been improved recently. After the addition of 4 more observing frequencies in the independent E-W and N-S arrays in 1992, it acquired the capability of instantaneous 2-dimensional mapping in 1996 (Kerdraon, 1997). A. Coulais (with J. L. Pincon, P. Lantos, A. Kerdaron and F. Lefeuvre) presented a new method for the computation of 2-D maps with rotational synthesis, based on tomography. It was applied in observations taken in May-June 1992 at 164, 236, 327 and 410 MHz and gives more reliable results than the classic CLEAN method (Fig. 5).

3 Dynamics and Heating

Although radio observations have given us important information about the structure of the solar atmosphere from the transition region up to the middle corona, they have not yet provided much information about its dynamics. This is due to instrumental limitations; in particular, aperture synthesis of regions with complex structure requires a good coverage of the u-v plane, which is not easy to achieve in snapshot observations, even with the VLA. Nevertheless the observations indicate time changes of the order of several minutes (Erskine and Kundu, 1982), which have been associated with the evolution of the network. Kundu et al. (1994a) found simultaneous flaring in soft X-rays and mm-λ in one out of four X-ray bright points that they detected at 17 GHz with the Nobeyama Radioheliograph; this was a long duration event, of apparently thermal nature. In the recent observations of A. Benz, time variations were almost simultaneous at 2 cm and in the soft X-rays.

The mere fact that we observe bright structures associated with the network or X-ray bright points is a clear indication of energy release. The question is whether this energy is sufficient to maintain and heat the corona. One may recognize different types of regions with different energy fluxes required to balance radiative and conductive losses (Withbroe and Noyes 1977, Withbroe 1988): for active regions and X-ray bright points the estimated energy flux required is $\epsilon \simeq 10^7$ erg/cm^2/sec, for the quiet corona $\epsilon \simeq 8 \times 10^5 - 10^6$ erg/cm^2/sec and for coronal holes $\epsilon = 5 \times 10^5 - 8 \times 10^5$ erg/cm^2/sec, including solar wind losses.

From optical observations it has been known for a long time (Beckers, 1972) that the matter lost by spicules is more than sufficient to balance the losses of the corona through the solar wind, while the kinetic energy of that matter is not enough to heat the corona. A. Benz estimated the energy associated with the bright structures at $\sim 10^{25}$ ergs/sec, which again is not sufficient to heat the corona. The fact that the energy released in the observed phenomena is not sufficient to balance the coronal energy losses has been debated in the working group, with the conclusion that it does not represent a surprise once the following theoretical scheme is accepted.

It is clear that there is more than enough energy present in the convection zone to supply total coronal losses. Energy is essentially injected from the photosphere as a Poynting flux $\mathbf{S} = c/4\pi(\delta\mathbf{E} \times \delta\mathbf{B})$, where $\delta\mathbf{E}$ and $\delta\mathbf{B}$ are the electric and magnetic fields induced by the photospheric motions perpendicular to the large scale magnetic field $\mathbf{B}_0$. Assuming the photosphere to be a perfect conductor, $\delta\mathbf{E} = -\delta\mathbf{v} \times \mathbf{B}_0/c$, whereas $\delta\mathbf{B} = \sqrt{4\pi\rho}\delta\mathbf{v}$. Adopting typical photospheric values for ρ, δv and B_0, the Poynting flux results of the order of 10^7 erg/cm^2/sec. The boundary motions of magnetic footpoints are due essentially to the solar granulation, with characteristic speeds $\delta v \simeq 0.25 - 2$ km/sec, sizes l_c of order $l_c \simeq 10^3$ km, and lifetimes τ_c of order $\tau_c \simeq 300$ secs, as well as the supergranulation, with characteristic speeds $\delta v \simeq 0.3$ km/sec, sizes $l_c \simeq 3 \times 10^4$ km and lifetimes $\tau_c \simeq 10^5$ seconds.

From the above arguments it follows that two major questions must be answered: how is magnetic energy supplied to the corona, and how is it dissipated?

The build-up of energy is easily understood, at least qualitatively, by recalling that the magnetic field lines threading the corona are rooted in the much denser photosphere, where they are continuously stressed by the local turbulent motions. Such stresses manifest themselves at coronal heights as induced current that enhance the large scale magnetic energy. The subsequent dissipation of the stored energy poses a severe problem since the collisional dissipative mechanisms acting in the corona are largely inefficient at the scales at which the magnetic energy accumulates. On the other hand, an MHD system like solar corona with very low values of the dissipative coefficients tends to be in a turbulent state under the action of an external forcing.

The properties of coronal turbulence, i.e. the way the energy can be stored in the magnetic field and then dissipated, are very difficult to model. In fact, in an active region modeled as a cube of side $L = 10^{10}$cm, the large scale magnetic Reynolds number for a field of 50 G, density $\sim 10^9$ cm^{-3} and temperature $\sim 2 \times 10^6$ K is $S \sim 10^{13}$. Assuming a homogeneous and stationary coronal turbulence, one finds that the Taylor microscale, defined as the energetically weighted average length over the entire inertial range, is $\lambda \sim S^{-1/2}L \sim 3 \times 10^3$ cm and is independent of the precise power–law spectrum provided the energy decreases with scale (Einaudi and Velli 1994a). The dissipative scale, at which the dissipation time scale equals the nonlinear time (a dimensional estimate of the time for the cascade toward small scales to occur), depends on the spectrum, $l \sim S^{-2/3}L \sim 20$ cm adopting the Kraichnan (1965) description of MHD turbulence. These numbers are of course indicative and represent orders of magnitude, but there is no doubt that in order to explain solar activity in terms of dissipation of magnetic energy on timescales of seconds or less, the magnetic field contributing to the available free energy must be structured over spatial scales of the order of one meter or less, where the local Reynolds number is of order unity. As a result the local release of magnetic energy occurs on the dynamical timescale and is concentrated inside current sheets which are continuously formed and dissipated throughout the system.

The idea that photospheric motions induce the formation of current sheets in the corona by displacing the magnetic field line footpoints has been proposed and studied in several ways in the past (Gold 1964; Parker 1972, 1983,1988,1991; Sturrock and Uchida 1981; Van Ballegooijen 1986; Mikic et al. 1989; Berger 1991; Einaudi and Velli 1994b; Longcope and Sudan 1994).

Einaudi et al. (1996a) have performed numerical simulations of a 2-D section of a coronal loop, subject to random magnetic forcing. These simulations last much longer than the coronal Alfvén time and the time scale of the driver in order to be able to follow in time the properties of the current sheets which continuously form and disrupt in the system. The forcing models the link between photospheric motions and energy injection in the corona. The boundary disturbances propagate along the vertical mean field B_0 with the associated Alfvèn velocity and give rise to a perpendicular magnetic field $B_\perp$. The appearance of $B_\perp$, and consequently of $v_\perp$, can be followed in time by solving the MHD equations in 2-D in which the terms coupling the perpendicular dynamics to the vertical are treated as imposed forcing terms. The results show the highly intermittent spatial distribution of current concentration generated by the coupling between internal dynamics and external forcing. In Fig. 6 we give an example of the current distribution as a function of position (below) and of the magnetic field lines contours (above). It is evident that the dissipation is concentrated in very localized current sheets separating large-scale magnetic loops (islands in two dimensions). The total power dissipation rate is a rapidly varying function of time, with jumps of orders of magnitude even at low Reynolds numbers, as shown in Fig. 7, where the spatial average of the current dissipation rate $< \eta J^2 >$ is plotted as a function of time.

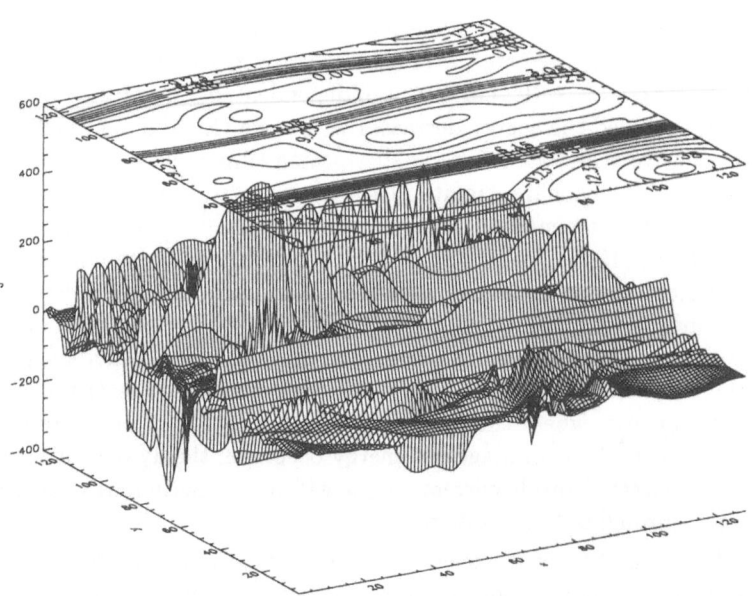

Fig. 6. A snapshot of spatial current distribution and magnetic field contours from 256×256 simulations performed by Einaudi et al. (1996a)

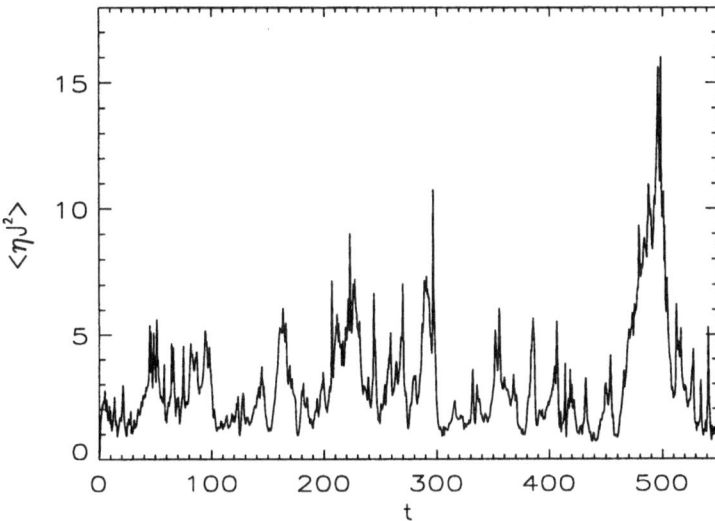

Fig. 7. Magnetic power dissipation rate in arbitrary units as a function of time in the $256x256$ simulations performed by Einaudi et al. (1996a)

Since both the spatial and temporal intermittency increases with the Reynolds number (resolution), it is reasonable to believe that current sheets are distributed all around the corona, continuously forming and disrupting here and there, releasing "elementary" bursts of energy whose intensities are in any case not detectable. When the dissipated power, integrated over a volume much bigger than the dimensions of the typical current sheet, jumps above the observational threshold, a release of energy, averaged over a large number of current sheets, is observed. Such "averaged" burst can involve energies over different orders of magnitude.

A very important question is whether we have processes of non-thermal energy release in the quiet sun. In the above theoretical scenario it is likely that strong electric fields are excited locally inside the current sheets and that at a scale larger than the collisional dissipative scale such electric fields are bigger than the Dreicer field, leading to runaway distribution functions. As a result the MHD approximation fails and the study of the final stage of the build up of the current sheets and their disruption must very likely be performed within the framework of kinetic theory. From a large scale point of point of view the "effective" Reynolds number in the corona due to such kinetic effects could be much smaller than the one derived using classical Spitzer resistivity.

The radio domain is quite appropriate for the detection of energetic particles that should be released in such processes. The fact that we have not detected them does not necessarily mean that they do not exist; it could very well be that we are faced with a very large number of small dynamic phenomena, mini or micro-flares, which produce a collective effect while they can not be detected individually due to our instrumental limitations. From this point of view, it will be very important to have imaging observations with time resolution of a few seconds and with measurement of circular polarization. Such observations should be able to detect impulsive events that might be associated with the appearance of bright radio features. Another possibility is to look for electron beams higher up in the corona, i.e. for type III bursts; Kundu et al. (1994b) have already identified type III's associated with flaring X-ray bright points.

Of particular interest is the observation presented by R. Lin (1977) in this conference, which showed the existence of a low density, high energy component in the interplanetary plasma which goes up to ~ 100 keV, the so called *super halo*. The origin of this component, which apparently is there all the time, is not clear. However the participants of the meeting, as well as the members of the working group, wondered whether this is a signature of energy release that heats the corona.

The effect of energy deposition in magnetic structures was studied by R. Walsh (with G. E. Bell and A. W. Hood), who presented computations of the response of the coronal plasma to a time varying heat source. In the case of periodically varying heating, they obtained critical timescales required in order to maintain a hot corona.

Finally, the role of the waves in the heating process of the corona was discussed on the basis of recent results (Einaudi et al. 1996b), showing that the waves are not necessarily required to carry *themselves* the energy needed to heat the corona, but may act as catalysts of nonlinear processes that give direct access to the energy stored in the large scale fields.

4 Conclusions

This is a section where we present more questions than real conclusions. A firm conclusion, however, is that the radio domain provides important diagnostics of the physical conditions, both in the case of low and the case of high spatial and temporal resolution. Radio models give a deficit of matter at low temperatures, with respect to models based on UV/EUV data. It is important to resolve this discrepancy and we have a very good opportunity in doing so: simultaneous observations with SOHO can provide the necessary data at high spatial resolution.

Different physical processes influence the emission of radiation in the two spectral domains, thus complementary diagnostics can be obtained by joint observations; moreover, radio is unique in providing information about the magnetic field. Joint observations can also give a better understanding of the difference in physical conditions between bright structures (network boundaries, bright points) and the darker cell interiors. They will also help us to understand better the dynamic behaviour of the transition region and the low corona. Statistics of the energy distribution among bright structures can also be performed, which may give some insight to the question of whether there is enough energy to heat the corona. Higher up, the Nançay Radioheliograph can now give information on the large scale dynamics, at a rate of 2 images/sec in five frequencies.

An important question is whether the physical processes responsible for the heating of the corona release energetic particles, since this information can help us identify the heating mechanism. The radio domain is very sensitive to radiation from accelerated electrons, however we have no evidence of non-thermal emission from the quiet sun so far. Imaging observations in the microwave range with time resolution of a few seconds and circular polarization measurements will help to identify impulsive, polarized radiation, which is the usual signature of non-thermal electrons. It will also be important to identify the origin of the "super halo" component detected in the interplanetary space.

Although there are many similarities in the physical processes taking place in active regions and the quiet sun, it is useful to keep the distinction between the two. The scale and the distribution of the magnetic field are different and so is the energy content and the rate of energy release. It is also an open question whether in quiet regions we have energy storage in the manner we have in active regions.

In comparing observations with theory, one has the feeling that the coupling between them has not yet reached the desired level yet, although we have had significant progress during the last years. We usually extrapolate from computations done at low Reynolds numbers into regimes where even MHD may not be valid. The observer would like to have concrete predictions of a particular theoretical model that can be checked with the available instrumentation, i.e. at a particular angular and temporal scale and in a particular frequency range. From this point of view, it will be important to develop a theoretical frame that does not only describe the formation of current sheets and reconnection, but also gives information on the heating and the release of energetic particles that will in turn be used to compute the characteristics of the electromagnetic radiation.

In this respect the turbulent nature of the corona can physically motivate statistical theories of solar activity such as the self-organized criticality (avalanche) model (Lu and Hamilton, 1991; Lu et al. 1993, Vlahos et al. 1995). The fundamental ingredient is the non-gaussianity of the response to a gaussian forcing which results from the locality of the instability criterion and of the relaxation process. A statistical analysis of the data resulting from the simulations at 256x256 resolution performed by Einaudi et al. (1996a) shows evidence of intermittent behaviour both in time and space, which means the existence of a non-gaussian tail in the distribution of the current around its mean (spatial or temporal) value. An analysis of the relationship between the properties of coronal turbulence and discrete avalanche models can probably lead to establish a physical background to these statistical theories which seem very promising in describing highly dynamical complex systems as solar corona.

One of the authors (G.E.) likes to thank M. Velli for discussions and a careful reading of the manuscript.

References

Alissandrakis C.E. (1994a): Adv. Space Res. **14**, (4)81

Alissandrakis C.E. (1994b): *Advances in Solar Physics*, (eds. G. Belvedere, M. Rodonò and G.M. Simnett), Lect. Notes Phys. **432**, 109

Alissandrakis C.E., Lantos P., Nikolaidis T. (1985): Solar Phys. **97**, 267

Alissandrakis C.E., Lantos P. (1996): Solar Phys. **165**, 61

Avrett E.H. (1995): *Infrared Tools for Solar Astrophysics: What's Next?*, (eds. J.R. Kuhn and M.J. Penn), Proc. of the 15th NSO/Sacramento Peak Workshop, p.298

Bastian T.S., Dulk G.A., Leblanc Y. (1996): ApJ *in press*

Beckers J.M. (1972): ARA&A **10**, 73

Benz A.O., Krucker S., Acton L.W., Bastian T. (1996): A&A *in press*

Berger M. A. (1991): A&A **252**, 369

Bogod V.M., Grebinskij A.S. (1996) Sov. Radiophys. (*in press*)

Bogod V.M., Korolkov D.V. (1975): Pisma AZh **1**, 25 (SvA Letters **1**, 205)

Borovik V.N., Kurbanov M.Sh., Makarov V.V. (1992): AZh **69**, 1288

Borovik V.N., Kurbanov M.Sh., Livshits M.A., Ryabov B.I. (1993): AZh **70**, 403

Borovik V.N., Kurbanov M.Sh., Livshits M.A., Ryabov B.I. (1989): AZh **67**, 1038 (SvA 34, 522)

Einaudi G. and Velli M. (1994a): *Advances in Solar Physics*, (eds. G. Belvedere, M. Rodonò and G.M. Simnett), Lect. Notes Phys. **432**, 149

Einaudi G. and Velli M. (1994b): Space Sci. Rev. **68**, 97

Einaudi G., Velli M., Politano H., Pouquet A. (1996a): ApJ **457**, L113

Einaudi G., Califano F., Chiuderi, C. (1996b): ApJ **472**, *in press*

Erskine F.T., Kundu, M.R. (1982): Solar Phys. **76**, 221

Fontenla, Avrett and Loeser (1993): ApJ **406**, 319

Gary D.E., Zirin H. (1988): ApJ **329**, 991

Gary D.E., Zirin H., Wang H. (1990): ApJ **355**, 321

Gold T. (1964): *The Physics of Solar Flares*, (ed. W. Hess), NASA SP-50, 389

Grebinskij A.S. (1987): SvA Letters **13**, 99

Habbal S., Roman R., Withbroe G.L., Shevgaonkar R.K., Kundu M.R.: 1986): ApJ **306**, 740

Habbal S.R., Gonzalez R.D. (1991): ApJ **376**, L25

Kerdaron A. (1997): This volume

Kundu M.R., Alissandrakis C.E. (1975): MNRAS **173**, 65.

Kundu M.R., Rao A.P., Erskine F.T., Bregman J.D. (1979): ApJ **232**, 1122

Kundu M.R., Shibasaki K., Enome S., Nitta N. (1994a): ApJ **431**, L155

Kundu M.R., Strong K.T., Pick M., Harvey K., Kane S., White S.M., Hudson H.S. (1994b): ApJ **427**, L59

Kundu M.R., Velusamy T. (1974): Solar Phys. **34**, 125

Lantos P., Alissandrakis C.E., Rigaud D. (1992): Solar Phys. **137**, 225

Lantos P., Alissandrakis C. E. (1996): Solar Phys. **165**, 83

Lin, R. P. (1997): This volume

Longcope L., Sudan, L. (1994): ApJ **437**, 491

Lu E.T., Hamilton, R.J. (1991): ApJ **380**, L89

Lu E.T., Hamilton, R.J., McTiernan J.M., Bromund, K.R. (1993): ApJ **412**, 841

Mikic Z., Schnack D.D., Van Hoven, G. (1989): ApJ**338**, 1148

Nitta N., Kundu, M.R. (1988): Solar Phys. **117**, 37

Parker E.N. (1972): ApJ **174**, 499

Parker E.N. (1983): ApJ **264**, 642

Parker E.N. (1988): ApJ **330**, 474

Parker E.N. (1991): ApJ **372**, 719

Raymond J.C., Doyle J.G. (1981): ApJ **247**, 686

Scudder J.D. (1992): ApJ **398**, 319

Sturrock P.A., Uchida, Y. (1981): ApJ **246**, 331

Vernazza J.E., Avrett E.H., Loeser R. (1981): ApJS **45**, 635

van Ballegooijen A.A. (1986): ApJ **331**, 1001

Vlahos L., Georgoulis M., Kluiving R. and Paschos P. (1995): A&A **299**, 897

Withbroe G., Noyes, R.W. (1977): ARA&A **15**, 363

Withbroe G. (1988): ApJ **325**, 442

Zirin H., Baumert B.M., Hurford G.J. (1991): ApJ **370**, 779.

Energy Release in the Solar Corona*

Timothy S. Bastian[1] and Loukas Vlahos[2]

[1] NRAO, P.O. Box 'O', Socorro, NM 87801, USA
[2] Department of Physics, Univeristy of Thessaloniki,
    54006 Thessaloniki, Greece

Abstract. Energy release in the solar corona drives a wide variety of phenomena, including flares, filament/prominence eruptions, coronal mass ejections, solar particle events, as well as coronal heating and the solar wind. The basic physics of these phenomena and their relationship to each other remains a vigorous area of inquiry. The *Working Group on Energy Release* at Mont Evray directed its attention to recent observational and theoretical developments relevant to flares and coronal heating. Particular attention was given to the "fragmentation" of energy release in solar flares and its interpretation; to the statistics of the flare phenomenon and whether they can be understood in terms of "driven dissipative systems"; to quasi-steady energy release and the problem of coronal heating; and to recent observations of flares and related phenomena.

Résumé. Les libérations d'énergie dans la couronne solaire sont à l'origine d'une grande variété de phénomènes incluant non seulement les éruptions, les éruptions de filaments/protubérances, les éjections de matière coronale, mais aussi le chauffage de la Couronne et le vent solaire. La compréhension de la physique de ces phénomènes et de la façon dont ils sont reliés entre eux est actuellement très lacunaire et constitue un domaine de recherche actif. Le groupe de travail sur la libération d'énergie, réuni à Mont Evray, s' est plus particulièrement intéressé aux observations et développements théoriques récents relatifs aux éruptions et au chauffage de la Couronne. Les discussions ont plus particulièrement portées sur: la fragmentation de l'énergie libérée et son interprétation; les propriétés statistiques des éruptions et leur interprétation possible en termes de systèmes dissipatifs forcés; les libérations d'énergie quasi-permanentes en liaison avec le problème du chauffage de la Couronne; des résultats nouveaux déduits de récentes observations des éruptions et de phénomènes associés.

1 Introduction

In this report we attempt to capture the content and, to some extent, the flavor of discussions held in the *Working Group on Energy Release* at the CESRA Workshop on *Coronal Physics from Radio and Space Observations* at Mont Evray. The issues which dominated discussions at Mont Evray were diverse, including the "fragmentation" of energy storage and release, whether such fragmentation might usefully be described in terms of "driven dissipative

---

* Working Group Report

systems", the problem of coronal heating, as well as observational work on flares.

In the interest of summarizing discussions in a reasonably coherent way, we make no attempt to present each of the contributions by individual group members in any detail. Rather, we have written a report which attempts to place the contributions and discussion within a general context which, we feel, reflects current trends on the issues raised. Accordingly, this report is organized as follows: in §2 we present a brief overview of energy release in the solar corona. In §3 we discuss the fragmentation of energy release. In §4 we discuss flare frequency distributions and models of energy release which treat active regions as driven dissipative systems. In §5 we discuss recent observational work. We conclude in §6.

2   Energy Release in Flares

Energy release in the solar corona produces a rich variety of phenomena: transient events such as flares, filament eruptions, and coronal mass ejections; and quasi-steady phenomena such as coronal heating and the solar wind. At Mont Evray, the *Working Group on Energy Release* was primarily concerned with aspects of the (possibly related) problems of flares and coronal heating.

At its root, energy release in flares involves the liberation of energy stored in magnetic fields or, equivalently, in the electrical currents. Two classes of theories of energy release in active regions have therefore most commonly been considered: those involving magnetic reconnection and those involving current dissipation. Melrose (1993, 1995) has considered the problem of energy release within both contexts. He points out that the two are formally equivalent, differing primarily in viewpoint rather than content, although at the practical level the differences are considerable. Theories involving magnetic reconnection are electrodynamically *local* in character, the energy released being that stored locally in the nonpotential part of the magnetic field. In contrast, theories involving current dissipation are electrodynamically *global* in character, in the sense that the entire current system is affected by the introduction of local dissipation, leading to the transport of energy from distant parts of the circuit to the dissipation region. Any successful theory of energy release in the corona must make sense from both viewpoints.

So, too, must any successful theory of the flare phenomenon account for the *spatial* and *temporal* scales on which energy is released, as well as the *dynamic range* of the energy release events. The largest releases of energy are solar flares and coronal mass ejections, both involving up to a few $\times 10^{32}$ ergs. The smallest releases of energy extend down to "microflares" (few $\times 10^{26}$ ergs) and perhaps even to "nanoflares" (few $\times 10^{23}$ ergs; see §4.3). Can flares be regarded as a superposition of "elementary bursts"? What factors determine whether a microflare or a large flare occurs in an active region? What determines the frequency distribution function of flare events? What determines the cutoffs in the size distribution of energy releases; i.e., is there an energy release quantum or quanta, and is there a maximum flare size?

These issues were central to discussions at Mont Evray. It is upon these issues, therefore, that we frame the next two sections of this report.

3   Fragmented Energy Release

The idea that energy release in flares involves numerous discrete events dates back more than two decades. Early analyses of Hard X-Ray (HXR) emission during the impulsive phase of flares found a great deal of structure on short timescales (seconds), structure which could be interpreted as a superposition of many events, leading van Beek et al. (1987) to coin the term "elementary flare burst". Similarly, Kaufmann et al. (1980, 1984) used the term "quasi-quantization" to describe energy release in flares on the basis of fine structure observed in mm-wavelength emissions. Kiplinger et al. (1983) found that HXR emission sometimes showed variations with rise times comparable to those seen in fine structures at mm wavelengths, of order 50 ms.

In more recent years a wealth of new observational data has become available. Fine structures in both the temporal and frequency domains have been observed in radio spectrograms (spike bursts, type IIIdm bursts; see review by Benz 1994). Aschwanden, Benz, and Schwartz (1993) and Aschwanden et al. (1995) have explored the correlation between hard X-ray and radio fine structures during flares and advocate a common cause for both: bi-directional electron beams. While it is extremely unlikely that *all* fine structure can be accounted for in terms of elementary bursts, the close correlation between decimetric and HXR fine structures strongly suggests that energy release in flares is fragmentary in some sense. Furthermore, there have now been many studies of X-ray, radio, and UV "microflares" or "transients" in active regions (Lin et al. 1984, Bastian 1991, Shimizu et al. 1992, Gopalswamy et al. 1994, White et al. 1995, Porter, Fontenla, and Simnett 1995, Gary, Hartl, and Shimizu 1996). From these studies it is now clear that energy release occurs over a large dynamic range – in "microflare" events up to the largest solar flares.

These observations have motivated theoretical work designed to explore the physics of fragmentary energy release. More than a decade ago Sturrock et al. (1984) advanced the idea that energy release in flares is fundamentally fragmentary in nature. Moreover, Sturrock et al. identified a hierarchy of timescales which characterize flare events: i) "sub-bursts" on timescales < 1 second; ii) "elementary bursts" on timescales of seconds; iii) a time scale of 10s of seconds, characteristic of the impulsive phase of energy release; iv) a timescales of tens of minutes, characteristic of the gradual phase. Sturrock et al. identify the various timescales with specific physical processes involved in energy release: "sub-bursts" are identified with the formation of magnetic islands as energy release proceeds via magnetic reconnection; an "elementary burst" is identified with the liberation of free energy in a single "elementary flux tube"; the impulsive phase is identified with the superposition of elementary bursts; and the gradual phase is identified with a period (before and after the impulsive phase) during which steady magnetic reconnection is presumed to occur, in contrast to "stochastic" reconnection during the impulsive phase.

During the intervening years, interest in fragmented energy release has become intense. Indeed, two recent workshops have been devoted to fragmented energy release and related issues (see van den Oord 1994, Benz and Krüger 1995). At Mont Evray, several contributions were relevant to the question of fragmented energy release. S. Krucker described new observations which demonstrated the occurence of fragmented energy release high in the corona. Specifically, Krucker and collaborators (see also Krucker et al. 1996, Benz, Csillaghy, and Aschwanden 1996) have performed joint spectroscopic and imaging observations of spike bursts at radio wavelengths. Krucker argued that the source of radio spikes was spatially coincident with the energy release volume, and that it was located on open magnetic field lines. The energy released in this volume was subsequently transported downward (producing a weak soft X-ray enhancement in the low-corona, releasing $\sim 10^{25}$ ergs) and upward (producing metric type III bursts).

On the theoretical side, considerable effort has gone into understanding the dynamic evolution of current sheets. L. Pustil'nik presented a model wherein energy release occurs in a turbulent reconnecting current sheet (RCS). He shows that magnetic reconnection cannot occur in a steady fashion. The current sheet has a tendency to "disintegrate" as a result of the tearing mode instability which tends to disrupt the sheet into current strings which, in turn, are subject to pinch instabilities. As the turbulent RCS fragments further, numerous clusters of "normal" and "anomalous" small scale resistive elements form leading Pustil'nik to speculate that current flow may be analogous to a percolation process through a random net of "good" and "bad" resistors (e.g., Redner 1983).

The free energy stored in current sheets may be liberated by a variety of MHD and plasma kinetic instabilities. These include tearing mode instabilities, coalescence, current-driven kinetic, and radiative instabilities. B. Kliem and J. Schumacher (see also Schumacher and Kliem 1996) presented numerical simulations of current sheet dynamics via 2D compressible MHD. Their model differed from that of Pustil'nik's to the extent that localized regions of anomolous resistivity were introduced explicitly to the RCS. The consequence of this was to produce induced tearing and a variety of dynamical effects including the formation of a plasmoid from the coalescence of small-scale current filaments and its subsequent ejection. The conversion of magnetic to kinetic energy proceeds in a highly irregular fashion with the timescale for energy release being ~ 60 ms and a spatial scale of ~ 50 km.

While energy release in flares is now widely believed to be fragmentary in nature, its interpretation is controversial. Benz (1994) echoes Sturrock et al. (1984) in suggesting a hierarchy of temporal scales associated with energy release. At Mont Evray, G. Huang and collaborators presented examples of microwave pulsations for which the energy content of each pulse was similar; they argue these are again indicative of "quasi-quantization" of the energy release process. U. Schwarz applied structure function analysis and wavelet analysis to perform a local decomposition of mm-λ emission from a solar flare. The result was a "scaleogram", a graphical representation indicating which temporal scales were prevalent during different phases of the flare. A number of timescales were evident although their relation to underlying physical processes was, as yet, unclear. The application of numerical techniques of this kind fall within a general effort (e.g., Kurths and Schwarz 1994, Isliker and Benz 1994) to exploit the tools of newly developed theory of complexity (fractal analysis) to address the important question of whether the rich temporal structure associated with transient solar phenomena – the fragmentation – is the result of deterministic nonlinear dynamics or stochastic processes. The *spatial* character of energy release remains largely unexplored in this context.

4   Flares as a Statistical Process

4.1   Flare Frequency Distributions

Hudson (1991) and Crosby, Aschwanden, and Dennis (1993) have summarized the extensive work done on the frequency distributions of various flare emissions and/or flare-related parameters. The frequency distributions of coronal radio emission, X-ray emission, interplanetary type III bursts, and interplanetary particle events have been computed. All can be represented as power-law distributions of the form $dN = Ax^{-\alpha}dx$, where dN is the number of events detected between x and $x + dx$ where x is some parameter of interest; i.e., the peak flux or count rate, the integrated flux or count rate, or the event duration. A and α are constants determined from fits to the data.

The flare frequency distributions based on HXR data have received the most attention. The existing data bases of HXR observations are large and relatively uniform in sensitivity (e.g., the SMM/HXRBS data base). HXR emission is optically thin – hence all of the emission is seen – and it is closely linked to the energy release during a flare if, as is widely believed, the HXR emission is dominated by thick-target emission from nonthermal electrons accelerated during the impulsive phase.

The results of studies of HXR events may be summarized as follows (Crosby et al. 1993, Lee, Petrosian, and McTiernan 1995): the flare frequency distribution formed as a function of peak flux of nonthermal electrons yields α in the range 1.5-1.7; the frequency distribution formed as a function of the total energy in nonthermal electrons yields $\alpha \approx 1.5$; the frequency distribution as a function of HXR flare duration yields $\alpha \approx 2$. The SMM/HXRBS data base yields power-laws over 2-3 decades of the relevant parameter (peak energy, total energy, duration). When these results are combined with HXR observations of microflares (Lin et al. 1984) the frequency distribution as a function of peak HXR flux is a power law over nearly six decades in energy (Crosby et al. 1993).

Bai (1993) has noted that while the frequency distributions are indeed well-fit by power laws, the index varies with solar cycle and perhaps with the 154 day period in flare recurrence. Using HXR data from the WATCH detector on board the GRANAT spacecraft, N. Crosby reported that the HXR flare frequency distribution as a function of peak count rate is harder than those obtained from the HXRBS data ($\alpha \approx 1.56$). Furthermore, she finds that if the events are binned according to event duration, the slope of the flare frequency as a function of peak count rate is steeper for short-duration events than it is for longer duration events. That is, for events with durations t_d less than 200 s, $\alpha \approx 2.13$; for $200 < t_d < 500$, $\alpha \approx 1.79$; for $500 < t_d < 800$, $\alpha \approx 1.54$; and for $t_d > 800$, $\alpha = 1.23$. If the flare frequency distribution is computed as a function of t_d, Crosby finds that the distribution is not well-fit by a single power-law. Instead, a double power-law is required, or a single power-law ($\alpha \approx 1.11$) with an exponential cutoff at a duration $t_d = 2394$ s.

Since SXR emission reflects the thermal energy content of a flare the frequency distribution of SXR flares is also interesting. Lee, Petrosian, and McTiernan (1995) compared the frequency distributions of SMM HXRBS and BCS events as a function of peak count rates with the frequency distribution of GOES SXR flares as a function of flare importance. They found that the slopes for all distributions were ≈ 1.8. Shimizu (1995) has examined a large number of active region SXR transient brightenings (Shimizu et al. 1992) with energies between $10^{25} - 10^{29}$ ergs. Although the results depend on the details of pixel averaging, he finds a result consistent with that of Lee et al. Shimizu also finds that the frequency distribution as a function of energy has an index $\alpha = 1.5 - 1.6$ for SXR transient energies $> 10^{27}$ ergs.

As pointed out by Lee et al. (1995) these are somewhat puzzling results. If the SXR emission is due to thermal plasma heated by nonthermal particles then the HXR fluence should be proportional to the peak SXR flux (the "Neupert effect"). Statistically speaking, the flare frequency distribution as a function of SXR *peak flux* should yield the same slope as the flare frequency distribution as a function of HXR *fluence*. Lee et al. find that, for an HXR-selected sample of flares, the frequency distribution as a function of HXR fluence yields a flatter slope ($\alpha \approx 1.6$). They conclude that these results are incompatible with the Neupert effect.

4.2   Driven Dissipative Models of the Flare Phenomenon

The statistics of flares in active regions have provoked a number of as yet unresolved questions. By what means can the flare frequency distribution function be understood? Can the entire range of flare energies – from microflares to large flares – be understood as superpositions of elementary bursts? If so, what ultimately defines the elementary burst quantum? Do elementary bursts occur stochastically? Or are they coupled in some fashion? Does such coupling lead to "self-organized" behavior? G. Haerendel pointed out that fragmentation may not be a matter of the energy release only – it may also apply to energy storage. In other words, a substantial part of the flare energy may be stored in small-scale structures. Attempts to understand the statistical properties of energy release in flares have led to a number of models. These have been constructed with varying degrees of physical content and sophistication.

One of the earliest attempts to account for the flare frequency distribution function was that of Rosner and Vaiana (1978) with what is essentially a "stochastic relaxation" model. Rosner and Vaiana assume that energy is built up exponentially with time in some volume; i.e., at a rate $dE_T/dt = \beta E_T$ where $E_T = E_0 + E$ is the total energy in the volume and E_0 represents the ground state energy. A "flare" is a stochastic process which releases the stored energy completely, returning the volume to the ground state E_0. If flare durations are short compared to the time between flares, then the probablility density of flaring (the probability that a flare occurs between time t and $t+dt$) is given by the Poisson distribution: $P(t) = \bar{\nu}e^{-\bar{\nu}t}$, where $\bar{\nu}$ is the mean flaring rate. Combining this with the equation for energy build-up in the flaring volume yields $p(E) = P[t(E)]|dE/dt|^{-1} = (\bar{\nu}/\beta E_0)(1 + E/E_0)^{-(1+\bar{\nu}/\beta)}$. One can then write the number of flare events per unit time which occur between E and $E + dE$ as $dN = A(1 + E/E_0)^{-\alpha}dE$ where $A = (\bar{\nu}^2/\alpha E_0)$ and $\alpha = (\bar{\nu} + \alpha)/\alpha$. For $E > E_0$, the frequency distribution approaches a power law.

Lu (1995c) has criticized Rosner and Vaiana's model on a number of grounds. Chief among them is that the prescription for energy storage is unlikely to be valid in solar active regions. The rate at which energy would be stored immediately following a flare is βE_0 while the storage rate immediately before the largest flares would be βE_{max}. The energy storage rate must then have a dynamic range comparable to the ratio of the largest flare energies to the smallest (observed to be $E_{max}/E_0 > 10^6$) and must be exponential over 14 e-foldings (or more). Since any deviation from an exponential energy storage law results in departures of the $\log N - \log E$ distribution from a power law, Lu regards this as unlikely.

Litvenenko (1994) has suggested a variant of the "stochastic relaxation" in which the problem of energy storage is perhaps placed on a more physical footing via reconnecting current sheets RCS. In the RCS model of energy storage and release, one or more physically independent RCS experience recurrent instabilities. Introducing a flare probability function $p(E) = dN/dE$ averaged over the flare-producing volume, Litvenenko begins with the assumption that $p(E)$ is subject to a time-independent "transport equation". Rather than assuming $dE/dT \propto E$ (Rosner and Vaiana 1978) Litvenenko employs the RCS model of Imshennik and Syrovatskii (1967) to show that the energy storage rate prior to an RCS event is $dE/dt \propto E^{7/4}$, allowing the transport equation to be recast and solved, yielding $p(E) \propto E^{-7/4} \exp[(E/E_1)^{-3/4}]$, where E_1 is a threshold energy defined in Litvenenko (1994). The frequency distribution is a power law with an index of $7/4$ for $E > E_1$; for energies $E < E_1$, however, the frequency distribution becomes much softer. More recently, Litvenenko (1996) has considered a RCS model wherein the dynamical evolution of single RCS or their interaction through coalescence is included. He finds that the flare frequency distribution function retains its power law character but, depending on the strength of a coalescence parameter, the index of the power law lies between $3/2$ and $7/4$, consistent with the range observed. The presence of a soft component in the distribution at small energies remains (see §4.3 below).

In contrast to the "stochastic relaxation" models described above, several workers have investigated cellular automaton (CA) models of solar flares. In this view, the fragmentation of energy release plays a fundamental role. Indeed, the underlying assumption is that flares are no more than a superposition of numerous elementary bursts. The strength of CA models is that the details of the physics of energy release and transport need not be included in order to gain insight into the statistics of the flare phenomenon. Typically the energy release mechanism is left unspecified in CA models, as is the means of communication between flaring elements since many candidates for flaring elements exist – e.g., RCS, current filaments, and/or double layers – and CA models are, in any case, insensitive to the details (MacKinnon, Macpherson, and Vlahos 1996). However, this strength may also be regarded as a weakness since, ultimately, any complete theory of the flare phenomenon must incorporate a detailed understanding of energy storage, release, and transport (see also Melrose 1995).

One class of CA models which has been explored extensively is the "sandpile" or "avalanche" model of flares. Lu and Hamilton (1991) first pointed out that the scale invariant (that is, power-law) nature of the flare frequency distribution could be described in terms of a system in a state of *self-organized criticality* (SOC; Bak, Tang, and Weisenfeld 1987). A common example of such a system is a sandpile to which sand is added until it reaches some critical slope. That is, it assumes a self-organized critical state. If more sand is added, the critical slope is exceeded locally and an avalanche occurs, bringing the slope back to its critical value. Avalanches of all sizes occur - from those involving one or a few grains up to those involving an entire side of the sandpile. The distribution of avalanche frequency with size is a power law. SOC systems are insensitive to initial conditions and require no fine tuning of parameters.

Lu and Hamilton first suggested that solar active regions were SOC systems, an idea subsequently developed by Lu et al. (1993), Lu (1995a,b), and by Vlahos et al. (1995). The general idea is implemented as follows: an active region is modeled as a vector field $\mathbf{F}$ on a uniform 3D lattice. The "slope" of the field at a given location i is defined as

$$d\mathbf{F}_i = \mathbf{F}_i - \sum_j w_j \mathbf{F}_{i+j} \qquad (1)$$

where w_j is a weighting function specified for a given model. One simple case involves only the six nearest neighbors; hence the slope involves a sum over six points weighted by w. The configuration is defined to be unstable at location i when the local slope exceeds some critical value: $|d\mathbf{F}_i| > \mathbf{F}_c$. When this occurs, the vector field is adjusted according to prescribed rules:

$$\mathbf{F}_i \rightarrow \mathbf{F}_i + \mathbf{f}_o, \quad \mathbf{F}_{i+j} \rightarrow \mathbf{F}_{i+j} + \mathbf{f}_j \qquad (2)$$

Here $\mathbf{f}_j$ are field vectors transported in the j direction defined so as to make $\mathbf{F}_i$ "stable" again ($|d\mathbf{F}_i| < \mathbf{F}_c$). However, by transporting the $\mathbf{f}_j$ from $\mathbf{F}_i$, neighboring lattice sites may be destabilized. An avalanche occurs, ending after all lattice sites once again satisfy the stability criterion. The rules for destabilization and transport can satisfy various conservation laws. For example, the field $\mathbf{F}$ is conserved by requiring $\sum_j \mathbf{f}_j = -\mathbf{f}_o$. The system is driven by adding vector elements $\delta\mathbf{F} \ll \mathbf{F}_c$ to random locations of the lattice. The slope at that site is computed after each addition of an element $\delta\mathbf{F}$. The addition continues until a site is destabilized and an avalanche occurs, after which the loading recommences.

In Lu and Hamilton, $\mathbf{F}$ is identified with the mean magnetic field and the slope at a given lattice site is defined to be the difference between the field at that site and the average field of its nearest neighbors (i.e., $w_j = 1/6$ for $|j - i| = 1$ and w=0, otherwise). The field diffusion to neighboring sites is affected by readjusting $\mathbf{F}_i$ according to $\mathbf{f}_j = \mathbf{F}_C/7$ if $|j - i| = 1$, $\mathbf{f}_j = -6\mathbf{F}_C/7$ if $|j - i| = 0$, and $\mathbf{f}_j = 0$, otherwise. Note that the direction of the $\mathbf{f}_j$ is given by $d\mathbf{F}_i/|d\mathbf{F}_i|$. Other rules are, of course, possible. Lu et al. (1993) found that while the system is insensitive to the choice of $\mathbf{F}_c$, it does depend on the choice of transport rules and the way in which the system is driven.

One might ask why *discretized*, SOC/CA models should be relevant to a *continuum* driven dissipative system such as a solar active region. Lu (1995a) has addressed this issue, arguing that a continuum system can indeed by described by a discrete CA model so long as the timescales governing the evolution of metastable states of the system, and timescales governing energy driving and release, meet certain conditions, and as long as local conservation laws apply to the field. Lu argues that solar active regions meet these conditions.

Among the successes of SOC/CA models (Vilmer 1996) are that they yield event frequency distributions as a function of peak energy release rate, total event energy, and event duration which are consistent with the observations. And, as pointed out by N. Crosby, the rolloff of flare frequency distributions at the largest energies is also accounted for as a consequence of the lattice

– the "active region" – being of finite spatial extent. Kucera et al. (1997) have pursued this aspect further, exploring the nature of the flare frequency distribution as a function of active region size. They again find results which may be accounted for within the context of SOC/CA models.

4.3   Quasi-Steady Energy Release

Energy release must occur in the Sun's outer atmosphere on a more or less continuous basis, as required by the existence of the corona and solar wind. There are other manifestations of quasi-steady energy release, perhaps related to coronal heating. One example is the presence of type I radio continuum and bursts – so-called storms – which are maintained for hours or days over solar active regions. V. Bogod and collaborators presented a comparison of the microwave emitting properties of active regions associated with noise storms observed by the RATAN-600 and the VLA (see also Bogod et al. 1995, 1996). They find that the polarization of the active regions associated with noise storms is complex: that transient polarization inversion is seen (for a day or so) and that the inversion is generally seen over relatively narrow bandwidths ($\approx$ 10%). They suggest that these features are perhaps best understood in terms of propagation through current sheets (e.g., Gopalswamy et al. 1994), and that the presence of current sheets may be related to the occurence of type I storms above active regions (e.g., due to emerging magnetic flux). If so, broadband microwave imaging may offer a means of determining the location of current sheets in the solar corona, the sites of quasi-steady energy release in the corona.

A second example was described by R. Lin at Mont Evray, and is reported in greater detail by Larson et al. (1996). The 3D Plasma and Energetic Particle experiment on board the WIND spacecraft has discovered a "super-halo" electron population in the interplanetary medium during periods free from solar energetic particl events and streams. This quiet-time super-halo is a nonthermal population of electrons with energies $\sim 2 - 100$ keV, believed to be of solar origin. Some energy release process in the corona must accelerate the super-halo component on a nearly continuous basis. Are distinct physical processes required to account for quasi-steady energy release? Or does quasi-steady energy release merely represent the high-frequency end of the flare frequency distribution, where extremely small releases of energy occur at a high rate? Or are a number of distinct energy release processes occuring, including extremely small flare events?

Parker (1988) suggested that "nanoflares" (events involving $\sim 10^{24}$ ergs or less) may indeed heat the X-ray corona. The fact that the flare frequency distribution function appears to be a power law over many decades led Hudson (1991) to consider the question of whether the smallest energy releases were likely to play a significant role in heating the corona. He pointed out that because the flare frequency distribution function is characterized by an index $\alpha < 2$ for event energies in the range of microflares up to the largest flares, the total power in the distribution is dominated by the events with the largest energies – hence the *observed* flare frequency distribution contains insufficient energy to heat the corona. In order for nanoflares to contribute significantly to coronal heating, the flare frequency distribution must possess a soft component at small event energies (i.e., $\alpha > 2$). Does such a component exist? If so, at what energy does the distribution index break from $\alpha < 2$ to $\alpha > 2$?

Both "stochastic relaxation" and SOC/CA models can be constructed which yield steep flare frequency distributions at small energies, i.e., in the nanoflare range of energies. As discussed in §4.2, the stochastic relaxation models of Litvenenko (1994, 1996) yield a soft frequency distribution on small energies. Vlahos et al. (1995) have explored models in which the transport rules are isotropic or anisotropic. They find that for the anisotropic case the instability criterion is more easily satisfied. Hence smaller, more numerous, events tend to occur, yielding a steep event frequency distribution. Vlahos et al. (1995) have therefore proposed a hybrid isotropic/anisotropic model which yields the observed frequency distribution function over an energy range corresponding to energies in the range of microflares to flares ($\alpha_1 \approx$ 1.8) and a steeper distribution ($\alpha_2 = 3.5$). More recently, Georgoulis and Vlahos (1996) have proposed an avalanche model wherein the loading is also modified. In particular, the loading is variable according to a probablility $P(\delta\mathbf{F}) = C(\delta\mathbf{F})^{-\kappa}$, where both C and κ are constants. A double power-law frequency distribution is again obtained, with a steep index for nanoflares ($\alpha_1 \approx 3.3$) and the observed index for energies in the microflare-to-flare range ($\alpha_2 \approx 1.7$). Note, however, that $\alpha_{1,2}$ depend on the assumed value of κ.

The observational evidence for such a steepening in the flare frequency distribution is largely absent (e.g., Lee, Petrosian and McTiernan 1993). At Mont Evray, however, C. Mercier and G. Trottet presented an intriguing analysis (Mercier and Trottet 1997) of 11 type I storms which may point toward a significant steepening of the frequency distribution function at extremely small energies (Fig. 1). In particular, they have formed the frequency distribution as a function of *peak flux density* of discrete type I bursts and have found that it is steep, with $\alpha \approx 3.0$, a value intermediate to indices resulting from the isotropic ($\alpha = 1.7$) and the anisotropic ($\alpha = 3.3$) models of Vlahos et al. (1995) and Georgoulis and Vlahos (1996). Note that the result is the same for type I bursts observed at both 164 and 237 MHz; a similar result is also obtained at 327 MHz albeit with poorer statistics. It is interesting to note that while the noise storms studied by Mercier and Trottet were associated with different active regions, the frequency distribution of each yielded approximately the same power-law index. Mercier and Trottet point out that if the peak flux density is correlated with the energy in nonthermal electrons, type I bursts represent the smallest discrete releases of energy observable

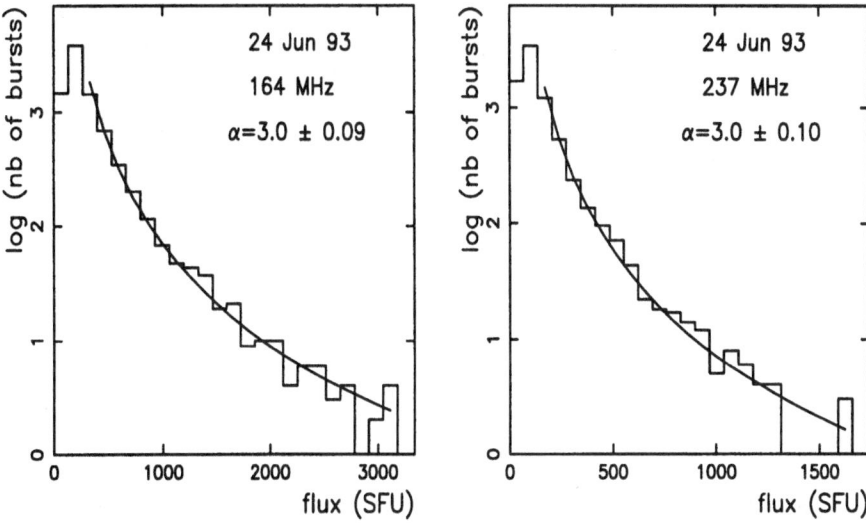

Fig. 1. The frequency distribution of type I bursts as a function of peak flux density at two different frequencies.

($\sim 10^{21}$ ergs; "picoflares"!). If so, they may offer insight into the general problem of heating active region loops.

This is a big "if", however. One must be cautious about interpreting the frequency distribution of type I bursts as a function of the underlying *energy released*. Type I bursts are believed to be the result of plasma radiation near the electron plasma frequency. Yet the details of the excitation of Langmuir waves by nonthermal electrons and their subsequent conversion to electromagnetic waves near the plasma frequency remains a poorly understood process. Furthermore, the relation of type I bursts to the general problem of coronal heating is uncertain - type I storms appear over some fraction of active regions for some fraction of their lifetime.

As a means of avoiding interpretive problems associated with type I bursts, it might be useful to pursue the approach employed by White et al. (1995) and Gary et al. (1996). These authors have shown that radio counterparts to soft X-ray transients (Shimizu 1992) are detectable as incoherent microwave events. Hence, sensitive interferometric observations at microwavelengths may yield insight into the nature of the the frequency distribution function at the smallest event energies.

However, it is useful to emphasize that in view of the results of Lee et al. (1995) and Shimizu (1995) described in §4.1, care must be exercised in interpreting estimates of the energy content of flares and microflares based on the observed radiation – HXR or microwave – resulting from nonthermal eletrons. If the efficiency with which electrons are accelerated depends on the size of the flare, or if the acceleration of any electrons at all requires a flare of some minimum size, the flare frequency distribution functions as a function of total energy inferred from nonthermal radiation may be in error.

4.4   The Flare Problem

The description of energy release in flares as a statistical process has yielded some insights into the character of energy release. Neverthless, the problem of understanding the discrete events – flares themselves – remains. For a flare involves more than the fact of catastrophic energy release; energy release must be placed within the larger context of energy storage and transport before the flare phenomenon as a whole can be understood.

How is energy stored in an active region? Is it "made" by driving magnetic footpoints via random photospheric fluid motions? Or are active regions "born", with well-developed current systems in place as they emerge from subphotospheric layers of the atmosphere (e.g., Leka et al. 1996). Is the energy storage itself "fragmentary"?

Given that energy release is derived from the free energy in magnetic fields (or, equivalently, currents) what conditions and mechanisms are involved in the conversion of that energy into plasma heating, particle acceleration, mass motions, and electromagnetic radiation? While reconnecting current sheets, current filaments, double layers, and DC electric fields all plausibly play a role, when, where, and under what circumstances are they relevant? How does the energy transport proceed in detail?

If coronal energy release is indeed fragmentary, how does the superposition of elementary flares yield largescale, sustained phenomena? For example, hard X-ray observations show the presence of well-defined macroscopic footpoints. Microwave imaging observations show the systematic evolution of spatially coherent coronal loops, as do soft X-ray observations. Are such structures simply the aftermath of fragmentary energy release, emerging as the result of energy transport processes (e.g., soft X-ray loops defined by locations where nonthermal beams drive chromospheric ablation, microwave loops defined by fast electrons in magnetic traps)?

How do CMEs and/or filament eruptions fit into the "statistical flare" scenario? CMEs appear to involve the largescale destabilization of the corona, leading to an ejection of plasma and magnetic field with an energy content often exceeding that of an associated flare, and CMEs are initiated *prior* to an associated flare. To be sure, CMEs are associated with fragmentary releases of energy, as is evident from fine structure present in associated radio emissions (e.g., Klein and Mann 1996) and white light coronagraphs. But can a CME be characterized, fundamentally, as a statistical process as flares have been, or do they represent a fundamentally different channel for energy release in the solar corona?

These questions indicate the continuing need for detailed and comprehensive studies of the many solar phenomena involving energy release in the solar corona: flares, CMEs, filament eruptions, solar particle events, coronal heating, and the solar wind. The relationship between these phenomena remains unclear, and our understanding of each is far from complete.

5   Observation and Interpretation of Flares

Previous sections have dealt with the general nature of energy release in flares and we pointed out the continuing need for the study of energy release events in their many guises. We now briefly turn to work on flares presented at Mont Evray. We do so from an observational point of view, where advances have been particularly rapid in recent years. The reasons for this are primarily instrumental. For example, spacebased mission such as *Yohkoh*, the *Compton Gamma Ray Observatory*, WIND, and *Ulysses* were all launched within the last five years. On the ground, the Nobeyama Radioheliograph was commissioned in 1992, and the capabilities at the Berkeley-Illinois-Maryland millimeter array (BIMA), the Owens Valley Solar Array, the Very Large Array, and the Nançay Radioheliograph have all been upgraded. New capabilities have become available through digititization of radio spectrographic (e.g., Zürich) and radiometric (e.g., Berne) observations.

5.1   Cm-λ/X-ray Observations

At Mont Evray, H. Nakajima presented multiband observations of an intense, extended flare, which occurred slightly behind the solar limb on 1992 Nov 2. The flare was observed with the *Yohkoh* soft- and hard-X-ray telescopes, and the Nobeyama 17 GHz radioheliograph. Nakjima found that the dominant 23-33 and 33-53 keV HXR sources were located in the soft X-ray high-temperature region, considerably higher than the corresponding soft X-ray bright loop. The 17 GHz microwave source, primarily due to electron with energies > 300, includes both the HXR and SXR sources (Fig. 2). Assuming a power-law electron energy distribution, radiometer measurements at 35 and 80 GHz implied a mean power-law index of ≈ 4. On the other hand, the hard X-ray spectrum derived from the HXT data shows a extremely soft spectrum with a power-law index of 7.2. Then, the electron spectrum index is estimated to be 6.7 or 7.7 depending on thick and thin target assumption, respectively. The time profiles of the hard X-ray and 17 GHz total emissions are similar to each other in overall structures.

Nakajima suggested that these results implied the presence of two distributions of electrons: a relatively soft distribution responsible for the HXR source, accelerated significantly above the SXR bright loops, and a harder distribution reponsible for the cm/mm-λ flux which is more pervasive. Nakajima argued that the source could be understood within the context of the model suggested by Tsuneta et al. (1992) wherein the energy release occurs

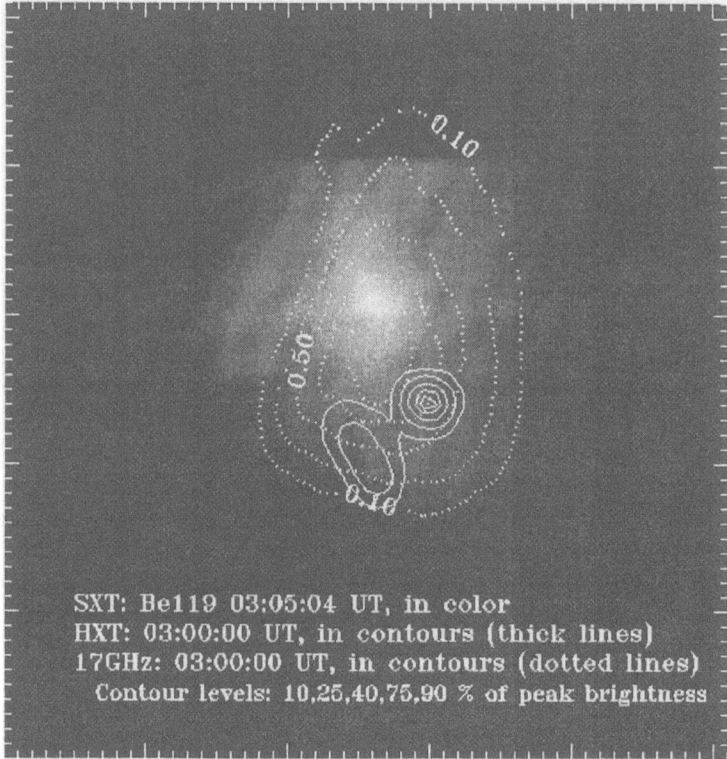

SXT: Be119 03:05:04 UT, in color
HXT: 03:00:00 UT, in contours (thick lines)
17GHz: 03:00:00 UT, in contours (dotted lines)
Contour levels: 10,25,40,75,90 % of peak brightness

Fig. 2. Overlay of microwave (17 GHz), hard X-ray (23-33 keV) and soft X-ray (Be119) images. The field of view is 157"x157". The solar north is up. The flare is estimated to have occurred beyond the west limb at (S23, W100).

via largescale magnetic reconnection above the SXR bright loops. Details will appear in Nakajima et al. (1996).

A second flare observed on the limb (1993 Jan 2) by the Nobeyama radioheliograph was discussed by K. Shibasaki. Shibasaki argued that mass motions played a significant role in the transport of energy in this event. The GOES M1.1 flare, lasting 2 hrs, was imaged at 17 GHz with a time resolution of 10 sec and an angular resolution of 20". Radiometer measurements between 2–17 GHz showed a flat flux spectrum, implying the emission was due to optically thin free-free radiation. The microwave source was initiated at one end of the flaring loop. Several fast-moving plasma clouds were produced which moved upward in the loop at a speed of 500-700 km s^{-1}. Shibasaki estimated each plasma cloud contained a kinetic energy of $\sim 10^{29}$ ergs and that the total kinetic energy contained in such clouds was $\approx 6 \times 10^{29}$ ergs, close to the total flare energy of 10^{30} ergs. The plasma clouds were stopped near the top of the flaring loop structure, growing brighter at 17 GHz as they were stopped, perhaps as a result of compression. The kinetic energy of the clouds was converted to thermal energy near the loop top. After the whole loop became bright (on a timescale of 15 min), a cusp like feature was formed from the bottom up, taking roughly 10 min to complete.

5.2   Mm-λ/X-ray Observations

Until recently most of the information on particle acceleration processes in solar flares was obtained from HXR and microwave observations. As a rule they provide information on electrons with energies $\lesssim 300$ keV. One of the interesting results obtained is that the majority of impulsive HXR bursts show a so-called "soft-hard-soft" behaviour whereas gradual and events with γ-ray line emission are characterized by a continuous hardening of the photon spectrum (Bai and Dennis 1985). There are two approches to explaining these features: i) they reflect the nature of the particle acceleration process (e.g., Bai 1986); ii) they are the result of the dynamics of energetic electrons in a flare loop (e.g., Vilmer, Kane, and Trottet 1982, Klein et al. 1986).

V. Melnikov and A. Magun attempted to undertand the temporal and spectral behaviour of the radio burst emission at cm- and mm-λ (8-50 GHz) using the second approach (see also Melnikov and Magun 1996). Using the data from the patrol instruments of the IAP (Bern University), they analyzed more than 20 impulsive and long duration radio bursts and have carried out calculations for the trap-plus-precipitation model of microwave emission (Melrose and Brown 1976).

The results of the analysis of 35 burst peaks with relatively simple time profiles are summarized as follows: i) The monotonic spectral hardening at frequencies well above the spectral maximum during the decay phase is typical for almost every well-pronounced peak in the flux time profiles; on the other hand, at frequencies close to the spectral maximum, a decrease of α becomes obvious (Fig. 3); ii) For the majority of the simple bursts a fast flattening occurs also during the rising phase; iii) time delays of the flux maxima at higher frequencies relative to those at lower frequencies appear in many events in the frequency range from our maximum frequency (50 GHz) down to the spectral maximum. The relative number of bursts with negative time delays increases in the two lower frequency ranges; iv) The spectral flattening and peak delays are common not only for long duration (gradual) but also for short impulsive bursts.

These results are in a good agreement with the expectations from dynamic trap models. Indeed, the continuous flattening of optically thin mm-spectra is consistent with the expected hardening of the energy spectrum of electrons as they are accumulated in a flare loop during injection. The delays of flux maxima at higher frequencies are expected from the fact that electron life times increase with energy and that in a trap their number reaches its maximum later. The detailed evolution of the spectrum at high and low frequencies is also accounted for – for example the variation of the spectral index at high

March 13, 1991

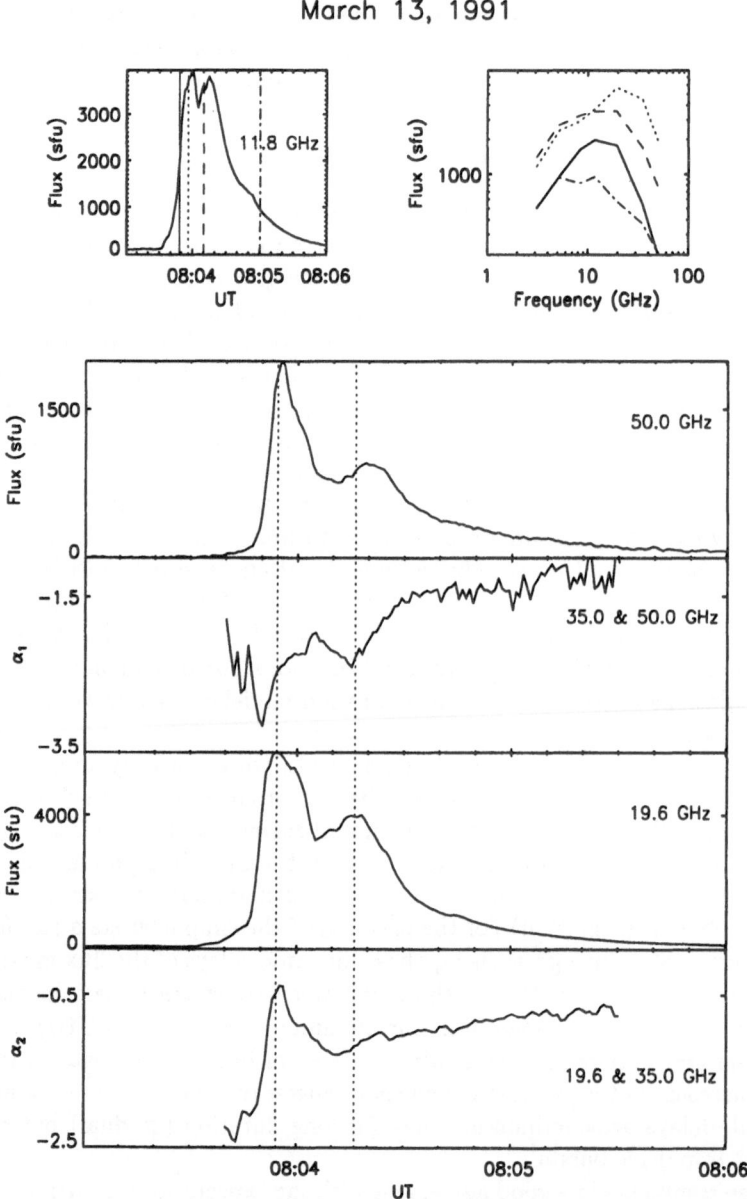

Fig. 3. Upper panels: The variation as a function of time of the 11.8 GHz microwave flux of a flare on 13 March 1991 (left); the integrated radio spectrum at times indicated by the dashed lines (right). Bottom panel: The variation of flux at 50 and 19.6 GHz and spectral indices computed from 35/50 GHz and for 19.6/35 GHz.

frequncies may be understood in terms of multiple injections of fast particles, while the evolution of the spectral index near the spectral maximum may be understood if the effects of time-dependent self-absorption are included.

In principle, the thermal bremsstralung emitted from an optically thin plasma, heated by fast electrons, could also produce spectral flattening (Chertok et al. 1995). However this becomes effective only in the late phase of bursts when a delayed thermal component could appear. Its typical time scale is much longer than the half intensity duration of the shorter bursts analysed by Melnikov and Magun, and they therefore feel justified in neglecting thermal bremsstrahlung emission.

S. Pohjolainen presented a study of seven flares observed with the 14 m antenna at Metsähovi at 37 GHz (see also Pohjolainen, Valtaoja, and Urpo 1996). These were compared with the GOES SXR and CGRO HXR counterparts. For the impulsive bursts (4/7) Pohjolainen found that the 37 GHz emission peaked 1-7 sec after the HXR maximum. She also found little evidence for significant thermal bremsstrahlung emission from these events. In contrast, the gradual bursts (3/7) showed evidence for significant thermal bremsstrahlung emission.

5.3   Energetic Particles

We refer the reader to Aschwanden and Treumann (1996) for a more comprehensive discussion of energetic electrons and ions in the solar corona and the interplanetary medium. As discussed briefly in §4.3, R.P. Lin and collaborators have identified a super-halo component of nonthermal electrons in the solar wind which they believe originate in the solar corona. P.K. Manoharan and colleagues have used the HISCALE instrument on board the *Ulysses* spacecraft to observe electrons in the 40–65 keV band when the spacecraft was at high latitiudes. Manoharan discussed an electron event detected at a helio-latitude of S74, an event which Pick et al. (1995) identified with a low-latitude flare. Using contemporaneous SXR images from the *Yohkoh* SXT, and data records from the Nançay radioheliograph, Manoharan argued that the event provided evidence for a large-scale reconfiguration of the coronal magnetic field followed by successive reconnection at increasing latitudes.

There has been recent interest in understanding which flares produce solar proton events (SPEs). Cliver et al. (1986) first noticed a tendency of flares associated with SPEs to show a progressive spectral hardening in HXRs (i.e., a "soft-hard-harder" spectral evolution). Kipinger (1995) has recently quantified this tendency, showing that those events displaying a soft-hard-harder spectral evolution of the HXR spectrum are likely to be associated with SPEs (for 82% of the events studied). Conversely, he found that events which did not show this evolution did not produce an SEP (for 95% of the events studied).

Q. Fu, G. Huang, Z. Gao, and Y. Liu presented an analysis of 12 microwave events at Mont Evray. The selected events were all SPEs for which radio observations were made at Bern. They note that the radio spectra of these events are more complex than the simple U-shaped spectrum of type IV-plus-microwave continuum, showing enhance emission at high frequencies with the spectrum more likely to be W-shaped. They find that the peak proton flux of an SPE is increases with the frequency of the spectral maximum at radio wavelengths.

Chertok et al. (1995) have studied the spectral flattening of radio bursts at mm-λ and have suggested a number of reasons for its presence: a super-position of self-absorbed gyrosynchrotron-emitting sources, a break in the underlying electron energy distribution function, a transition from gyrosyn-chrotron radiation in microwaves to synchrotron radiation at mm-λ, and a component emitting optically thin thermal free-free radiation.

Melnikov and Magun noted that variation of the radio spectra they stud-ied are very similar for both long duration (gradual) and short duration (impulsive) bursts. Photon spectra of hard X-ray bursts on the other hand show a more varied behaviour (Bai and Dennis, 1985). The majority of impul-sive hard X-ray bursts show a so called "soft-hard-soft" behaviour whereas gradual and events with γ-ray line emission are characterized by a continuous hardening of the photon spectrum. The latter is consistent with the flattening of mm-λ spectra if one assumes a hardening of the common electron spectrum. The discrepancy between the spectral behaviour of impulsive hard X-ray and microwave bursts can be due to the fact that the emissions originate from electrons in different energy ranges with different spectral evolution. Before final conclusions can be made further studies are needed in which hard X-ray and γ-ray spectra are compared with those of the microwave emission from the same events.

6 Summary and Questions

The temporal and spectral fragmentation of radio and HXR emissions during solar flares has now been abundantly illustrated to the extent that most flares show fine structure in their radio emission, some of which is closely correlated with HXR fine structures. The idea that the energy release itself is *fundamentally* fragmented, not only in flares, but in other manifestations of solar (and, by extension, stellar) activity, is a possibility which dominated discussions in the CESRA working group on energy release.

A second issue, which was raised repeatedly, was this: if energy release is indeed fundamentally fragmentary in all of its manifestions on the Sun (and stars), why is it that many phenomena – CMEs, soft-X-ray emitting loops, hard-X-ray emitting footpoints – appear to be *spatially* well-organized. Can a fragmentary or "statisitical" model satisfactorily account for the CME phenomenon, for example? This issue, in turn, raised a third point: by what means, observationally, can the hypothesis that the energy release in flares and other forms of solar activity (e.g., coronal heating) is fragmentary be placed on a more solid footing? By what means, observationally, can one distinguish between various models of energy release?

In a number of recent theoretical studies the dynamics of entire active regions has been studied. They proceed with the idea that energy storage and/or release could be described in terms of complex dynamical systems. The main assumption in this approach is that the evolution of the entire active region is not sensitive to the details of local processes. Using a variety of cellular automaton models these theories have been able to reproduce the flare frequency distripution as a function of maximum event flux, energy, and duration. The global models used so far are still very preliminary. Combining 3-D MHD simulations with CA models will be a very intresting research tool for the next generation of global active region models.

A number of observational directions can and, doubtless, will be pursued. For addressing the question of energy release in flares, it is important to understand the differences in slope of flare distribution functions based on events selected according to differing criteria – are the differences in detail real? Can they be understood within the context of CA/SOC models? Can we reconcile distributions based on estimates of the nonthermal and thermal energy content (e.g., Lee et al. 1995)? A promising approach is that outlined by Crosby and by Kucera et al. (1996), where the distributions are considered within the context of the properties of the active regions in which the flares occur.

There are the flares themselves. The possibility that energy release in flares is fragmentary is largely based on radio spectrograms, HXR light curves, and comparisons of the two. Radio, HXR, and SXR imaging have not been exploited to address this question. From the standpoint of modeling active regions as complex, driven, dissipative systems, it is important to analyze the spatial, as well as the temporal, properties of these emissions. Where do the elementary bursts occur within an evolving active region? How does an "avalanche" manifest itself in these emissions?

Answering the important question of whether coronal heating is the result of elementary flare bursts ("nanoflares") and if so, whether such nanoflares are really the low-energy extension of the much larger energy release events we call flares, depends on establishing the nature of the smallest energy releases in the solar corona. Observations of the kind reported by Lin may be of critical importance to this question, as may be the observations reported by Mercier and Trottet (1997), or by Gary et al. (1996).

The question of whether energy storage is itself fragmentary will require careful analyses of the birth and evolution of solar active regions (e.g., Leka et al. 1996) using vector magnetograms obtained with a high degree of spatial and temporal resolution, in conjunction with high resolution SXR images of the corona, as well as an analysis of when and where flares occur in the active regions under study.

These topics remain as challenges for future CESRA meetings.

Acknowledgement We thank the members of the CESRA Working Group on Energy Release for their participation in the discussions which served as the basis of this report. They include R. Bentley, V. Bogod, F. Chiuderi-Drago, N. Crosby, Q. Fu, G. Haerendel, B. Kliem, S. Krucker, R. P. Lin, A. Magun, P. K. Manoharan, V. Melnikov, C. Mercier, H. Nakajima, S. Po-hjolainen, L. Pustil'nik, J. P. Raulin, U. Schwarz, K. Shibasaki, G. Simnett, G. Trottet, and N. Vilmer. We also thank the CESRA scientific and local organizing committees for a stimulating meeting.

References

Aschwanden, M.J., Benz, A.O., and Schwartz, R.A. (1993): ApJ, **417**, 790

Aschwanden, M.J., Montello, M., Dennis, B.R., and Benz, A.O. (1995): ApJ, **440**, 394

Aschwanden, M.J., and Treumann, R.A. (1996): this volume

Bai, T. (1986): ApJ, **308**, 912

Bai, T. (1993): ApJ, **404**, 805

Bai, T. and Dennis, B. (1985): ApJ, **292**, 699.

Bastian, T.S. (1991): ApJ, **370**, L49

Bak, P., Tang, C., and Weisenfeld, K. (1987): Phys. Rev. Lett., **59**, 381

Benz, A.O. (1994): in *Sp Sci Rev*, **68**,

Benz, A.O., Csillaghy, A., and Aschwanden, M.J. (1996): A&A, **309**, 291

Benz, A.O. and Krüger, A. (eds.) (1995):

Bogod V.M., Garaimov V.I., Gelfreikh G.B., Lang, K.R., Willson R.F., and Kile J.N. (1995): Solar Physics, **160**, 133.

Bogod V.M., Garaimov V.I., Gelfreikh G.B., Willson R.F., Lang, K.R., and Kile J.N. (1996): Solar Physics, to be submitted.

Crosby, N., Aschwanden, M., and Dennis, B.R. (1993): Solar Phys, **143**, 275

Chertok, I.M., Fomichev V.V., Gorgutsa, R.V., Hildebrandt, J., Krüger, A., Magun, A., Zaitsev, V.V.: 1995, Solar Phys, **160**, 181

Cliver, E.W., et al. (1986): ApJ, 305, 920

Einaudi, G., and Alissandrakis, C. (1996): this volume

Gary, D.E., Hartl, M., and Shimizu, T. (1996): ApJ, submitted.

Georgoulis, M.K., and Vlahos, L. (1996): ApJ, 469, L135.

Gopalswamy, N., et al. (1994): ApJ, **437**, 522

Gopalswamy, N., Zheleznyakov, V.V., White, S.M., and Kundu, M.R. (1995): Solar Phys, **155**, 339

Hudson, H. (1991): Solar Phys, **133**, 357

Imshennik, V.S., and Syrovatskii, S.I. (1967): *Soviet Phys. JETP*, **25**, 656.

Isliker, H., and Benz, A.O. (1994): in *Fragmented Energy Release in Sun and Stars*, G.H.J. van den Oord (ed.), Kluwer:Dordrecht, p. 185

Kaufmann, P., Strauss, F.M., Opher, J., and Laporte, C. (1980): A&A, **87**, 58.

Kaufmann, P., Correia, E., Costa, J.E.R., Dennis, B.R., Hurford, G.J., and Brown, J.L. (1984): Solar Phys, **91**, 359

Kiplinger, A.L., Dennis, B.R., Emslie, A.G., Frost, K.J., and Orwig, L.E. (1983): ApJ, **265**, L99

Kiplinger, A.L. (1995): ApJ, **453**, 973

Klein, K.-L., Trottet, G., Magun, A. (1986): Solar Phys, **104**, 243.

Klein, K.L., and Mann, G. (1996): this volume

Kucera, T.A., Dennis, B.R., Schwartz, R.A., and Shaw, D. 1997, ApJ, accepted.

Kurths, J., and Schwarz, U. (1994): in *Fragmented Energy Release in Sun and Stars*, G.H.J. van den Oord (ed.), Kluwer:Dordrecht, p. 171

Krucker, S., Benz, A.O., and Aschwanden, M. (1996): A&A, submitted.

Larson, D., et al. (1996): in *Solwind 8 Conf. Proc.*, M. Neugebauer (ed.), Dana Point, CA, June 1995, in press

Lee, T.T., Petrosian, V., and McTiernan, J.M. (1993), ApJ, **412**, 401

Lee, T.T., Petrosian, V., and McTiernan, J.M. (1995), ApJ, **448**, 915

Leka, K.D., Canfield, R.C., McClymont, A.N., and van Driel-Gesztelyi, L. (1996): ApJ, **460**, 1019

Li, C.S. and Fu, Q.J. (1995): Acta Astrophys Sin, **15**, 350

Lu, E. (1995a): *Phys. Rev. Lett.*, **74**, 2511

Lu, E. (1995b): ApJ, **446**, L109

Lu, E. (1995c): ApJ, **447**, 416

Lu, E., and Hamilton, R.J. (1991): ApJ, **380**, L89

Lu, E., Hamilton, R.J., McTiernan, J.M., and Bromund, K.R. (1993): *Astrophys. J.*, **412**, 841

Litvinenko, Yu. B. (1993): Solar Phys, **151**, 195

Litvinenko, Yu. B. (1996): Solar Phys, submitted.

Lin, R.P., Schwarz, R.A., Kane, S.R., Pelling, R.M., and Hurley, K.C. (1984): ApJ, **283**, 421

Mercier, C., and Trottet, G. (1997): ApJ (Letters) **474**.

Melrose, D.B., and Brown, J.C. (1976): MNRAS, **176**, 15

Melrose, D.B. (1993): Aust J Phys, **46**, 167

Melrose, D.B. (1995): ApJ, **451**, 391

Parker, E.N. (1988): ApJ, **330**, 474

Pohjolainen, S., Valtaoja, E., and Urpo, S. (1996): A&A, **306**, 973

Porter, J.G., Fontenla, J.M., and Simnett, G.M. (1995): ApJ, **438**, 472

MacKinnon, A.L., Macpherson, K.P., and Vlahos, L. (1996): A&A, **310**, L9

Masuda,S., Kosugi,T., Hara,H., Tsuneta,S., and Ogawara,Y. (1994): Nat, **371**, 495.

Melnikov, V., and Magun, A. (1996): Solar Phys., submitted.

Nakajima, H., Fujiki, K., Metcalf, T.R., Kane, S.R., and Akioka, M. (1996): Solar Phys, to be submitted

Redner, S. (1983): Ann Israel Phys Soc, **5**, 447

Raadu, M.A. (1994): in Fragmented Energy Release in Sun and Stars, G.H.J. van den Oord (ed.), Kluwer:Dordrecht, p. 29

Rosner, R. and Vaiana, G.S. (1978): ApJ,**222**, 1104

Schumacher, J., and Kliem, B. (1996): Phys. Plasmas, submitted

Shimizu, T., Tsuneta, S., Acton, L.,W., Lemen, J.R., and Uchida, Y. (1992): PASJ, **44**, L147

Shimizu, T. (1995): PASJ, **47**, 251

Sturrock, P.A., Kaufmann P., Moore P.L. and Smith D.F. (1984): Solar Phys., **94**, 341

Tsuneta, S., Hara, H., Shimizu, T. Acton, L. W., Strong, K. T., Hudson, H. S., and Ogawara, Y. 1992, PASJ, **44**, L63.

van Beek, H.F., de Feiter, L.D., and de Jager, C. (1974): Space Res., **14**, 447 van den Oord, G.H.J. (ed.) (1994): Fragmented Energy Release in Sun and Stars, Dordrecht:Kluwer, 378 pp

Vilmer, N. (1996): this volume

Vilmer N., Kane, S.R., and Trottet, G. (1982): A&A, **108**, 306

Vlahos, L., Georgoulis, M., Kluiving, R., and Paschos, P. (1995): *Astron. Astrophys.*, **299**, 897

Wentzel, D.G., and Seiden, P.E. (1992): ApJ, **390**, 280

White, S.M., Kundu, M.R., Shimizu, T., Shibasaki, K., and Enome, S. (1995): ApJ, **450**, 435

Part II

Particle Beams

Observations of the 3D Distributions of Thermal to Near-Relativistic Electrons in the Interplanetary Medium by the Wind Spacecraft

Robert P. Lin

Space Sciences Laboratory and Physics Department, University of California, Berkeley, California, 94720-7450 USA

Abstract. Recent electron observations by the 3-D Plasma and Energetic Particles experiment on the WIND spacecraft are reviewed. Impulsive solar electron events observed at 1 AU are found to have power-law spectra often extending down to $\lesssim 1$ keV, implying a source region at 0.2-$1R_\odot$ altitude. At solar quiet times, a nonthermal, "super-halo" component extending from ~ 2 to $\gtrsim 10^2$ keV is found to be always present. For impulsive events which produce solar type III radio emission, the full 3-D distribution function of the electrons is obtained. During times of strong Langmuir wave bursts, the reduce, parallel distributions $f_r(v_\parallel)$ often are plateaued or slightly unstable, with $\frac{\partial f_r(v_\parallel)}{\partial v_\parallel} > 10^{-27}$, implying growth rates of $\gamma \sim 0.03$ s^{-1}. Most of the time, however, the reduced distributions are stable, with negative slopes, suggesting that other competing wave processes such as whistler growth may be stabilizing the beam.

Résumé. Nous présentons une revue des observations d'électrons obtenues récemment avec l'expérience "3-D Plasma and Energetic Particles" embarquée sur le satellite WIND. Ces observations montrent que les événements à électrons impulsifs, d'origine solaire, présentent un spectre en loi de puissance qui s'étend jusqu'à des énergies $\lesssim 1$ keV. Ceci implique que la région d'accélération est située à des altitudes comprises entre 0.2 et $1R_\odot$. Pendant les périodes de Soleil calme, on détecte en permanence une composante non-thermique, dite "super-halo", dans un domaine d'énergie allant d'environ ~ 2 à $\gtrsim 10^2$ keV. La fonction de distribution 3-D des électrons a été déterminée pour les événements impulsifs qui produisent des sursauts radio de type III. Durant les périodes de forte activité d'ondes de Langmuir les distributions parallèles réduites $f_r(v_\parallel)$ presentent souvent un plateau ou sont faiblement instables, avec $\frac{\partial f_r(v_\parallel)}{\partial v_\parallel} > 10^{-27}$, ce qui implique des taux de croissance $\gamma \sim 0.03$ s^{-1}. Cependant, les fonctions de distibutions réduites sont le plus souvent stables. Ceci suggère que le faisceau d'électrons doit être stabilisé par d'autres processus, par exemple par la croissance des siffleurs.

1 Introduction

The steady-state solar wind electron population is dominated by a core with temperature $kT \sim 10$ eV, containing $\geq 95\%$ of the plasma density and moving at about the solar wind bulk velocity, plus $\sim 5\%$ in a hot, $kT \sim 80$ eV, halo population carrying heat flux outward from the Sun, often in the form of highly collimated *strahl* [*Feldman et al.*, 1975]. At energies of $\sim$ keV and above, impulsively accelerated solar electron events occur at the Sun, on average, several times a day or more during solar maximum. As these electrons escape they produce solar and interplanetary type III radio bursts through beam-plasma interactions [see *Lin*, 1990 for review].

In the absence of these impulsive electron bursts, streams of energetic electrons have been observed in the interplanetary medium from storms of hectometric type III bursts (Bougeret et al., 1983). Here we review the first observations of electrons at 1 A.U. spanning the entire energy range from solar wind thermal plasma, $\lesssim 10$ eV, to near-relativistic energies ($\gtrsim 300$ keV), obtained by the 3-D Plasma and Energetic Particles Experiment (Lin et al., 1995) on the Wind spacecraft, launched in late 1994. For impulsive events, the electron spectrum is found to extend down to $\sim$0.5 keV, implying an acceleration source height of $\sim$0.2 - $1R_\odot$. At quiet times a "super-halo" component is detected, extending in a power-law from $\sim$1-2 keV to $\sim 10^2$ keV, suggesting that non-thermal processes may be operating continually in the solar corona. For the impulsive events which generate type III emissions, detailed 3-D distribution of electrons and plasma wave measurements are presented and compared to theoretical expectations.

2   Experimental Details

The Wind 3-D Plasma and Energetic Particle experiment consists of solid-state telescopes (SST) and electron and ion electrostatic analyzers (EESA, PESA). The SST consists of three arrays of semiconductor detectors, each with a pair of double-ended telescopes to measure electron and ion fluxes above $\sim$20 keV. EESA-L and -H are a pair of electrostatic analyzers with very different geometric factors to cover the wide range of electron fluxes from $\sim$3 eV to 30 keV and provide significant measurements even at the lowest flux levels likely to be encountered. Details of the detectors and experiment electronics are given in Lin et al. [1995].

The WIND spacecraft was launched in November 1994 into a double lunar swingby orbit. The observations presented here were obtained when WIND was in the interplanetary medium at geocentric distances greater than $\sim$60 R_E, well in front of the Earth's bow shock.

3   Solar Impulsive Electron Events

Figure 1 shows the first observation [Lin et al., 1996] of solar impulsive electron events spanning the entire energy range from solar wind to suprathermal particle (few eV to hundreds of keV). A solar impulsive electron event begins at 1100 UT, easily identified by its velocity dispersion, e.g., the faster

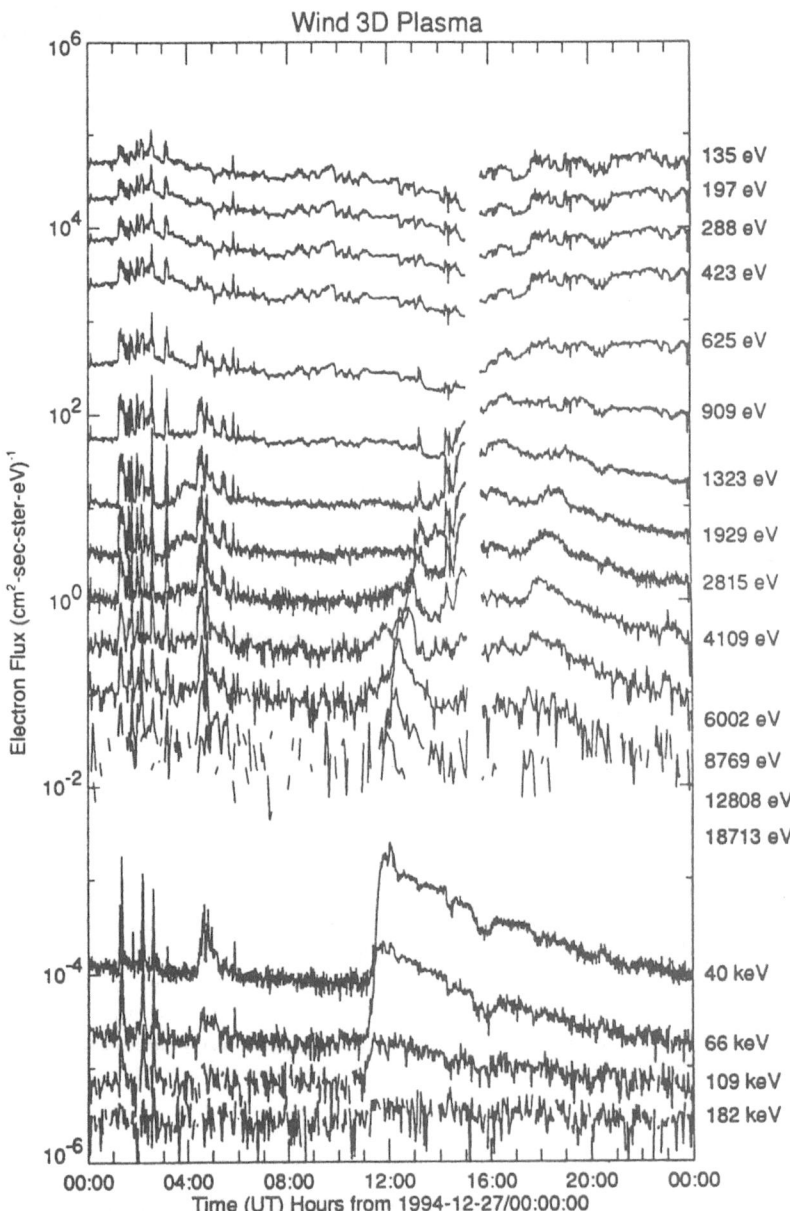

Fig. 1. Electron fluxes from ~100 eV to $\gtrsim$ 100 keV for 27 December 1994. The solar electron event begins at ~1100 UT at ~100 keV, with velocity dispersion evident down to 624 eV. Two smaller events, at ~1100 UT and 1620 UT, are visible below ~6 keV. The dip at ~1500 UT at low energies is due to the close approach to the Moon, resulting in a plasma shadow. (from Lin et al. 1996)

electrons arriving earlier, as expected if the electrons of all energies were simultaneously accelerated at the Sun and traveled the same distance along the interplanetary field to reach the spacecraft. The solar event can clearly be identified down to the 0.908 keV and even the 0.624 keV channels. A second, much weaker impulsive electron event is seen beginning about 1620 UT in the 8.77 keV channel. Another very small event may be starting at ~1120 UT at ~6 keV energy. The 3-D angular distributions in Fig.2 show

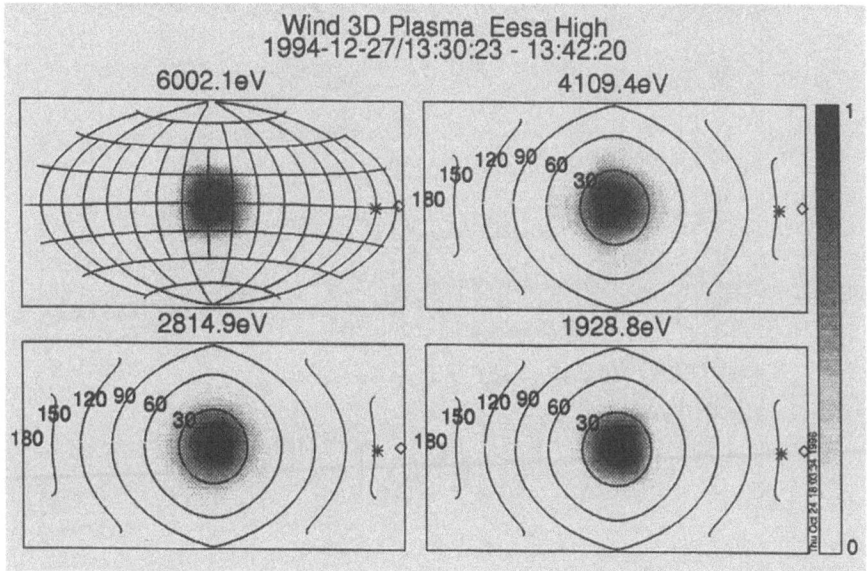

Fig. 2. The 3-D angular distribution of the electrons during the 27 December 1994 impulsive event. Each plot represents the normalized flux on a surface of constant energy in the solar wind plasma rest frame. The Hammer-Aitoff equal area projection is used to display 4π steradians angular coverage. The magnetic field direction is centered at the origin and the Sun direction indicated by the asterisk. The angular grid of the actual measurements is shown in the upper left panel, and contours of constant pitch angle (numbers) are shown in the other panels. (from Lin et al. 1996)

that the electrons in the main event are streaming outward from the Sun, with pitch-angle distribution highly peaked, within $\lesssim 30°$, along the magnetic field. Fig.3 shows electron differential flux v. energy spectra at different times during, and integrated over the entire impulsive event, as well as the pre-event solar wind spectrum, with core, halo and "super-halo" (discussed later). The event spectra have the pre-event spectra subtracted. The peak in the primary event spectra progresses to lower energies with time, although contributions from the smaller events are also evident. The event-integrated

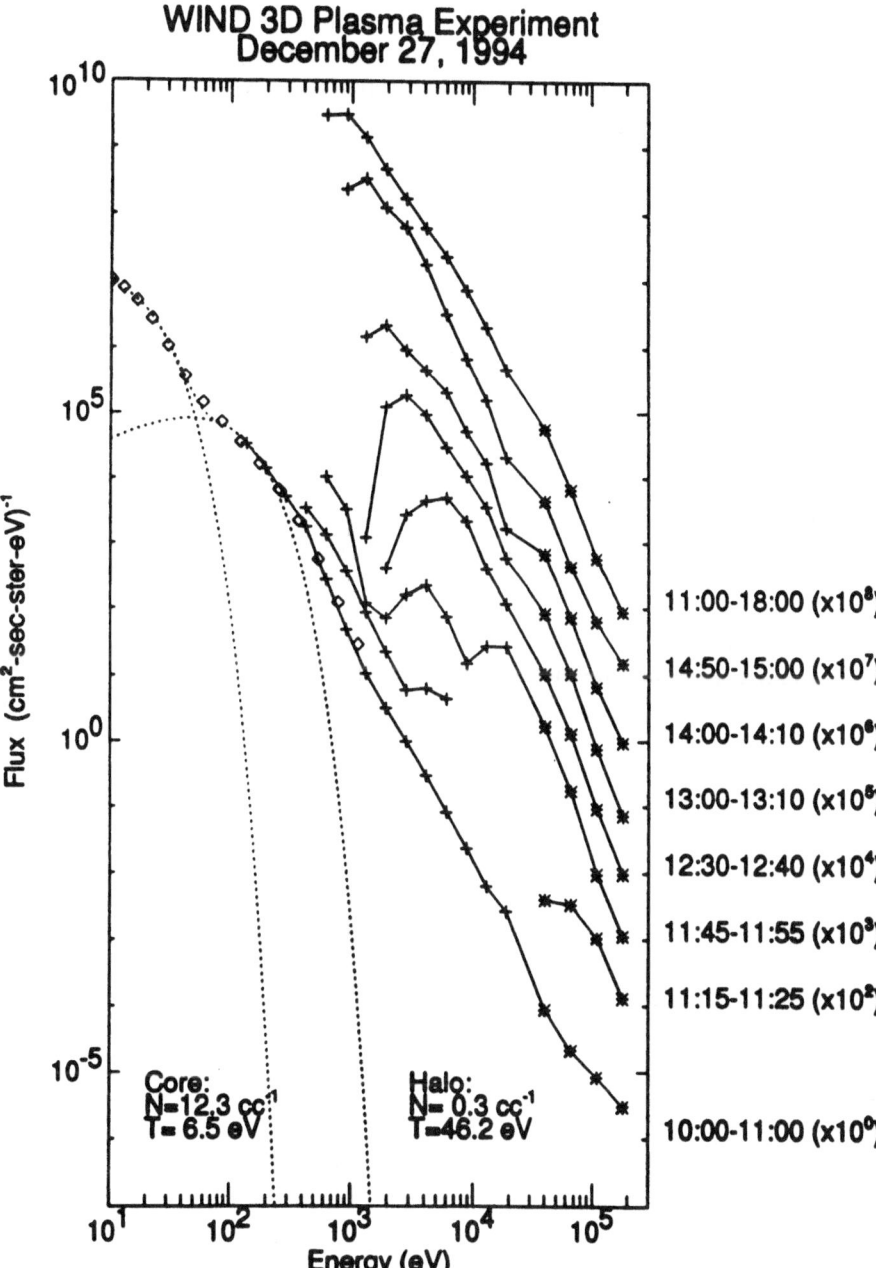

Fig. 3. Omnidirectional electron spectra (with pre-event electron fluxes subtracted) are shown for various times during the 27 December 1994 event, and averaged over the entire event. The pre-event electron spectrum shows the solar wind electron core and halo components, as well as a "super halo" extending to $\gtrsim 100$ keV. (from Lin et al. 1996)

spectrum displays a peak at $\lesssim 1$ keV, with significant flux at ~ 0.5 keV. No electrons are detected at 422 eV or below for this impulsive event. Above the peak, the spectrum is similar to those reported previously above ~ 2 keV [Potter et al., 1980], and can be fit to a power-law shape $dJ/dE = AE^{-\delta}$, where E is the electron energy in keV and A and δ are constants. The best fit gives $\delta = 3.0$ from ~ 1 keV to 40 keV, steepening to $\delta = 4.4$ above ~ 40 keV.

Because at coronal temperatures electrons are not gravitationally bound while protons are, an ambipolar electric field (the Pannekoek-Rosseland field [Pannekoek, 1922; Rosseland, 1924]) is set up with a total potential drop of about 1 kV from the base of the corona to 1 AU. This potential varies inversely with distance from Sun center, and it accelerates protons outward and decelerates electrons. Thus, the peak in the spectrum of the electrons just as they escape the corona would be up to ~ 1 keV more than measured at 1 AU, e.g., ranging from $\lesssim 1$ up to ~ 2 keV for the event of Fig.1 depending on the height of the acceleration.

The fact that the event spectrum extends down to such low energies indicates that at least some of the electron acceleration must occur high in the corona, since the range of $\sim$ keV energy electrons in ionized hydrogen, due to Coulomb collisions, is short compared to the column depth through the corona. Assuming that the initial accelerated electron spectrum is power-law with the same exponent as seen at energies above the peak, the maximum overlying column density can be calculated [Lin 1974]. For a peak at ~ 1.5 keV, the column density must be less than $\sim 9 \times 10^{17}$ cm^{-2}. This value implies that the lowest energy electrons must have been accelerated at altitudes of $\sim 1 R_{\odot}$, for the typical active coronal density models derived from radio observations [Dulk and McLean, 1978], or $\sim 0.2 R_{\odot}$ for the quiet equatorial corona at sunspot minimum [Saito et al., 1977].

Even though the first year of the WIND observation, late 1994 to late 1995, is near solar minimum, tens of impulsive solar electron events have been seen, with many detected down to ~ 0.5 keV. The coronal flare acceleration process thus appears to produce a power-law spectrum extending down to $\lesssim$ 1.5 keV or lower, compared to a coronal thermal electron energy of $kT \approx 0.1$ keV ($\sim 10^6$ K). Integrating over energy spectrum and over the duration of the event, and assuming that the cone of propagation for the electron event of Fig.3 is ~ 400, we estimate a total energy of $\gtrsim 3 \times 10^{26}$ ergs in escaping electrons. Thus, at least that much energy was released in the coronal flare process at the Sun.

4   Quiet Time Electrons

Unlike ISEE-3, the WIND spacecraft was launched near the minimum in the solar activity cycle. Thus, there are substantial periods free from energetic solar particle events and streams. Figure 4 shows the Wind omnidirectional

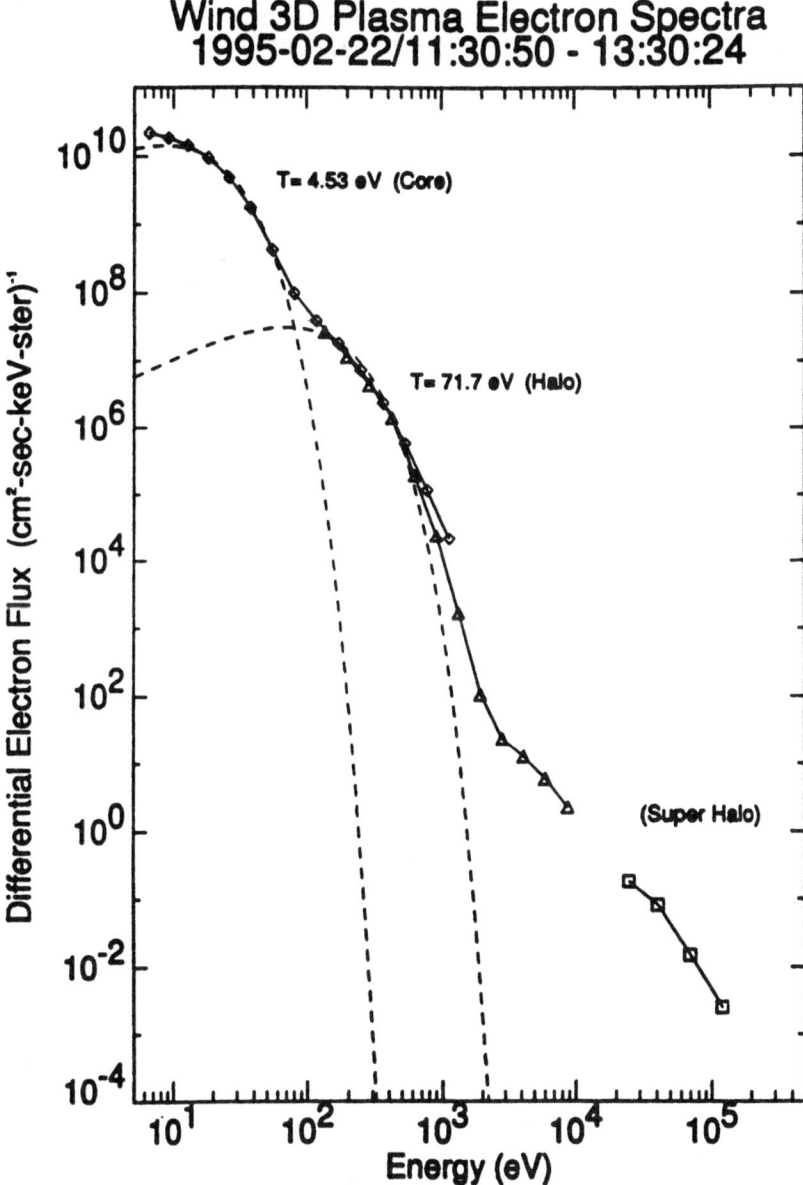

Fig. 4. Electron differential flux spectrum from ~5 eV to $\gtrsim 100$ keV, measured at a very quiet time, in the absence of any solar particle events, by the Wind 3D Plasma and Energetic Particle experiment (Larson et al., 1996). The diamonds, triangles, and squares indicate the three different detectors used to accommodate the wide range of fluxes over this energy range. The dashed lines give fits to Maxwellians for the solar wind core and halo.

electron spectrum from ~5 eV to ~ 10^2 keV measured during such a quiet period on February 22, 1995. Because of the extremely wide range of electron fluxes, three separate detectors–EESA-L, EESA-H, and SST–are required to measure this spectrum. The solar wind plasma Maxwellian core dominates from ~5 to ~50 eV; the solar wind halo takes over from ~ 10^2 eV to ~ 1 keV. The halo is believed to be due to the escape of coronal thermal electrons which have a temperature of ~ 10^6 K. Note, however, that the halo spectrum departs significantly from a simple isothermal Maxwellian at energies above ~0.7 keV.

A third, much harder component is evident beginning above ~2 keV and extends to $\gtrsim 10^2$ keV, which we have denoted the "super-halo" [Larson et al., 1996]. The spectrum of the "super-halo" appears to be approximately power-law with exponent $\delta \approx 2.5$. If this "super-halo" is solar in origin, (a more detailed analysis will be required to confirm this), it would imply that electrons of such energies must be continuously present at and escaping from the Sun. Of particular interest here, is the likely possibility that such a non-thermal electron population at the Sun may be detectable through its radio emission in the metric and decametric ranges.

It should be noted here that for exospheric models of the solar wind (Maksimovic et al., 1996) the presence of a significant non-thermal tail to the coronal electron population is sufficient by itself to accelerate the solar wind. The mechanism for producing this non-thermal population is unknown, but it is tempting to speculate that it is related to the mechanism which heats the corona (see, for example, Scudder, 1992). Further analysis and correlations with solar coronal observations are required to resolve the origin of these particles and their relationship to coronal heating.

5   Energetic Electron and Plasma Wave Observations, April 2, 1995

Four solar impulsive electron events ocurred on April 2, 1995 (Ergun et al., 1996). Figure 5a plots the omnidirectional electron fluxes at 96 s resolution as measured by EESAH. The center energies of the channels range from 140 eV to 27.1 keV, with the lowest energy traces on top. Immediately below (Fig.5b) are the omnidirectional electron fluxes as measured by the SST-Foil detector. The impulsive electron events can be clearly seen in the electron fluxes, with the strongest beginning at ~1110 UT with the 182 keV electrons and extending down in energy to ~600 eV.

A persistent feature in all of the events were ~15 minute period modulations of the electron fluxes (see the top panel of Fig.5) that appeared as the low-energy (~1 keV to ~4 keV) electrons arrived. Similar flux modulations can be seen in several other solar impulsive electron events observed by the WIND satellite and by the ISEE 3 satellite (see Lin et al., 1981, Fig.3). These modulations showed no measurable velocity dispersion, and the modulations of the second event are accompanied by modulations in the magnitude of the magnetic field that had a similar period, suggesting a hydromagnetic instability.

Figure 5c-d are spectrograms of the ratio (F/F_0) of the event electron fluxes (F) to the pre-event flux (F_0), determined by averaging from ~00:00 UT to ~04:00 UT. Superimposed on the spectrograms are linear fits of the arrival time of the electrons versus their inverse velocity. In all of the events, velocity dispersion of the electrons is consistent with the expected Archimedean spiral length (~1.15 AU) for the measured solar wind velocity (~370 km/s).

All four of the solar impulsive electron events were associated with solar type III radio bursts (Fig.5e-g) observed by the Wind Waves instrument (Bougeret et al., 1995). The RAD2 panel (Fig.5e) displays the radio emissions from 1 MHz to 14 MHz, RAD1 panel (Fig.5f) displays the 20 kHz to ~1 MHz emissions, and the TNR (Thermal Noise Receiver) panel (FIg. 5g) has high resolution in the frequency range from 4 kHz to 250 kHz including the local plasma frequency.

Each of the events had the typical signature of a solar type III radio burst. The electromagnetic emissions during events 3 and 4 extended up to 14 MHz, the maximum frequency of the Waves experiment. The lowest frequency electromagnetic emissions appeared to have a cut-off near the local second harmonic.

The narrow-banded emissions that persisted through out the day varying between 20 kHz and 40 kHz (see Fig.5g) are locally generated Langmuir waves; the power in Langmuir emissions is shown in Fig.5h. The series of Langmuir bursts between ~11:45 UT and ~12:45 UT coincide with the arrival of ~12 keV through ~2 keV electron fluxes of event 3. The bursts at ~15:25 UT coincided with the arrival of ~6 keV electron fluxes of event 4. The bursts at ~17:15 do not appear to be related to the solar impulsive electron events, but rather to a possible magnetic connection with the Earth's foreshock.

6   Electron Distributions

Figure 6a is an expanded view of Fig.5h between 11:45 UT and 12:40 UT. The nonlocal solar type III radio emissions contributed to the background power of $\sim \mu$V/m. The local Langmuir bursts rise 1 to 2 orders of magnitude above the background, reaching $\sim 50\mu$V/m. For this interval, the full 3-D electron distributions (see Fig.2 for example) averaged over 96 sec, were reduced to 2-D distributions (Fig.7b) by assuming gyrotropy. These 2-D distributions showed that the electron beam angular width was narrow, typically $\sim 16°$ at 6 keV, similar to that reported in the ISEE 3 observations (Lin et al., 1981). The 2-D distributions were then integrated over $v_\perp$ to obtain the reduced, one-

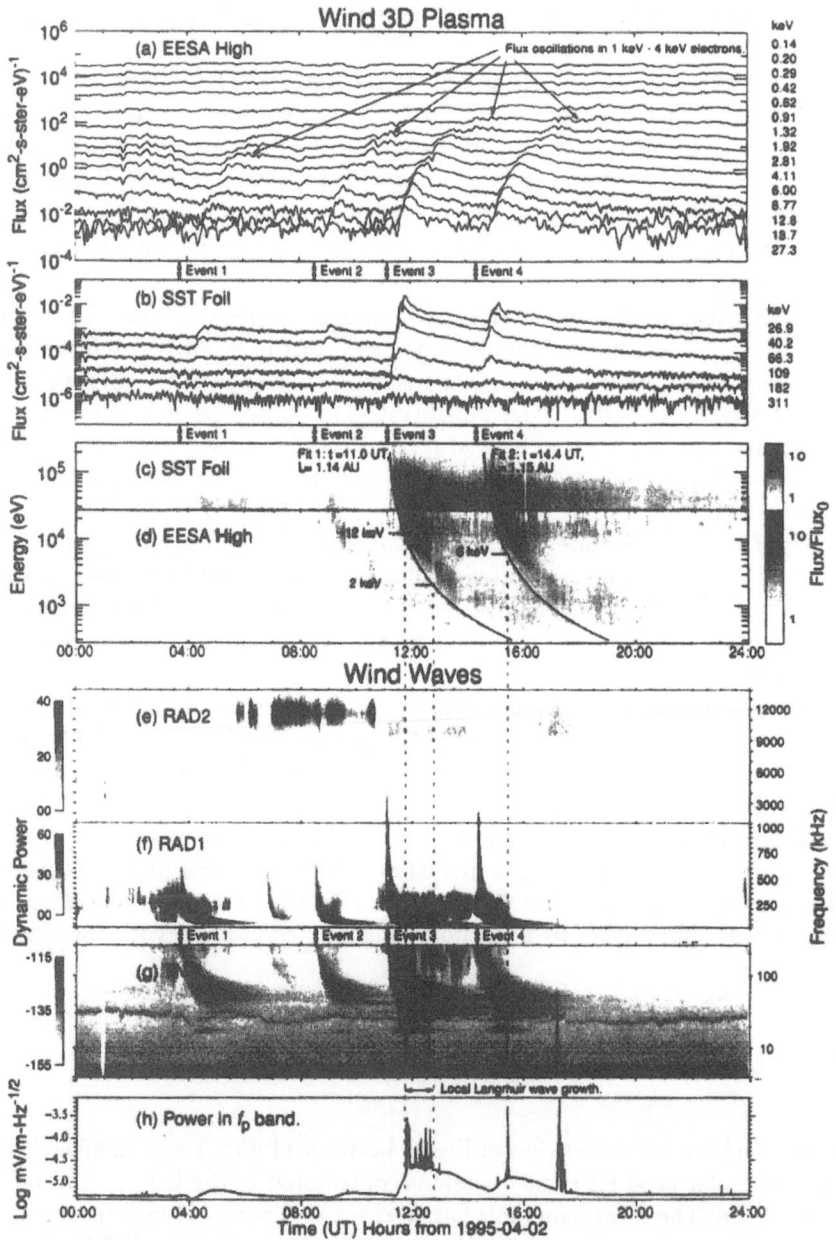

Fig. 5. (previous page) (a) The omnidirectional electron fluxes at 96 s resolution as measured by EESAH. The center energies are listed on the right. The fifteen minute period flux oscillations are marked. (b) The omnidirectional electron fluxes as measured by the SST-Foil detector. Again, the center energies are listed on the right. (c) and (d) Spectrograms of the ratio (F/F_0) of electron fluxes (F) to a reference flux (F_0) determined by averaging the electron fluxes over the period $\sim$00:00 UT to $\sim$04:00 UT prior to the first event. The vertical axis represents energy while the darkness of shade represents electron flux. Superimposed on the plot are linear fits of the arrival time of the electrons versus their inverse velocity. The event start times and path lengths as determined by the linear fit are marked on the plot. (e), (f), and (g) Solar type III radio bursts as observed by the Wind Waves instrument (Bougeret et al., 1995). The RAD2 panel displays the radio emissions from 1 MHz to 14 MHz (linear frequency axis) with the darkness of shade representing power relative to cosmic background (logarithmic scale) that spans approximately a factor of thirty. The TNR (Thermal Noise Receiver) panel displays the frequency range from 4 kHz to 250 kHz. The grey scale is in absolute units of $\log(\text{mV/m - Hz}^{-1/2})$ as measured by antennae with $\sim$50 m effective baseline. (h) The electric field wave power in the frequency band (19 kHz to 41.5 kHz) encompassing the local Langmuir frequency (from Ergun et al. 1996).

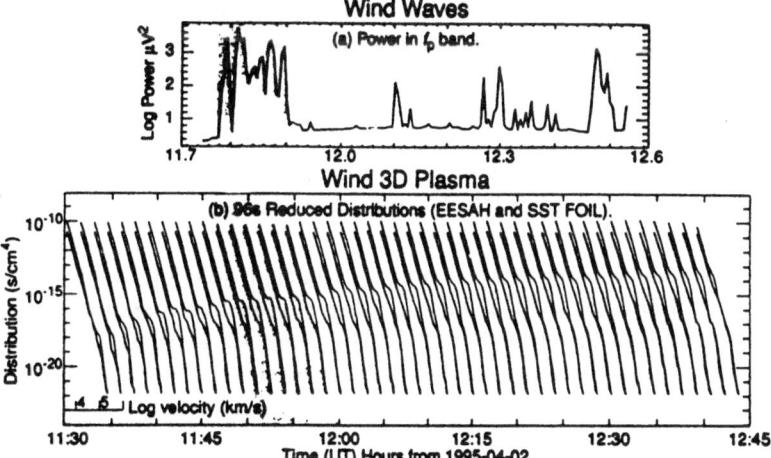

Fig. 6. (a) An expanded view of the wave power near the local plasma frequency for April 2, 1995 (see Fig. 5). The sharp peaks generally indicate local Langmuir emissions. The strongest series of wave emissions occurred from $\sim$11:47 to $\sim$11:55 UT (shaded). Below (b) are a series of 96 s reduced distribution functions on a log-log scale plotted in pairs. The top curve of each pair represents the $+\mathbf{B}$ direction, the bottom plot the $-\mathbf{B}$ direction. The vertical axis is the distribution function. The horizontal axis represents both velocity and time. The electron distributions are positioned such that the lowest velocity point ($\sim 7 \times 10^3$ km/s) indicates the beginning time of the 96 s averaging. A velocity scale for the first distribution is located below left. During the time of strongest wave emissions (shaded), the distribution functions were unstable or marginally stable (plateaued) while the later distributions were clearly stable. (from Ergun et al. 1996)

dimensional, electron distribution functions $f(v_{\parallel})$ every 96 seconds (Fig.7a).

These distributions every 96 s are plotted in Fig.6b on a log-log scale in pairs. The horizontal axis represents both time and velocity, the velocity axis for the first distribution is immediately below. The electron distributions are positioned such that the lowest velocity point indicates the beginning time of the 96 second averaging period. The upper trace of a pair represents the $+\mathbf{B}$ direction which includes anti-sunward traveling electrons, while the lower trace is the $-\mathbf{B}$ direction. At low velocities, the gap between the upper and lower traces reflects the anisotropic halo electrons which carry heat flow away from the Sun. At higher velocities ($> 10^4$ km/s), the electrons are more isotropic, so the traces generally lie nearly on top of each other. The separation between the two traces here is due to the solar impulsive electron event. Even in the leftmost distribution, the impulsive electron event was already underway.

During the Langmuir wave bursts from $\sim$11:47 UT to $\sim$11:55 UT, highlighted in grey in Fig.6, the reduced electron distributions appear to have been a marginally stable, or slightly unstable, with the beam velocity at $\sim 6 \times 10^4$ km/s (10 keV). Throughout the rest of the period the reduced electron distributions were stable. The observed positive slopes (see Fig.7a) at that parallel velocity are only marginally significant with $\partial f_r / \partial v_{\parallel} \gtrsim 10^{-27}$ $s^2 cm^{-5}$, corresponding to a growth rate of $\sim$0.03 s^{-1}. This growth rate is far less than the maximum growth rate of $\sim$0.1 s^{-1} to $\sim$1 s^{-1} seen by ISEE 3. The ISEE 3 also had several events which had positive slopes the endured for long ($>$10 min) periods. Such strong, persisting positive slopes have not yet been seen by Wind.

The reduced, one-dimensional distribution functions were almost always stable ($\partial f_r / \partial v_{\parallel} < 0$). If kinetic relaxation from Langmuir waves dominated the electron evolution, the distributions are expected to have been plateaued, especially during periods of Langmuir emission. Velocity dispersion and adiabatic focusing (traveling into region of lower $|B|$) should continually act to destabilize the distributions. It is significant, then, that most of the distributions were stable with negative slopes. These observations indicate that, in addition to quasilinear relaxation, other processes are contributing to the evolution of the electron beam.

Even when a reduced, parallel distribution $f_r(v_{\parallel})$ is plateaued or stable, a slice (not reduced) of the three-dimensional distribution along $\bar{B}$ may have a positive slope. Such a distribution is stable to Langmuir waves and other parallel waves, but still has free energy for oblique electrostatic waves or electromagnetic waves. Interestingly, slices of the three-dimensional distribution function along B were often plateaued, and $\sim$15% of the time (between 11:30 UT and 12:45 UT) displayed a positive slope. One possibility is that kinetic nonlinearities from oblique waves may play a role in electron evolution. Whistler waves have been reported to be associated with Langmuir emission

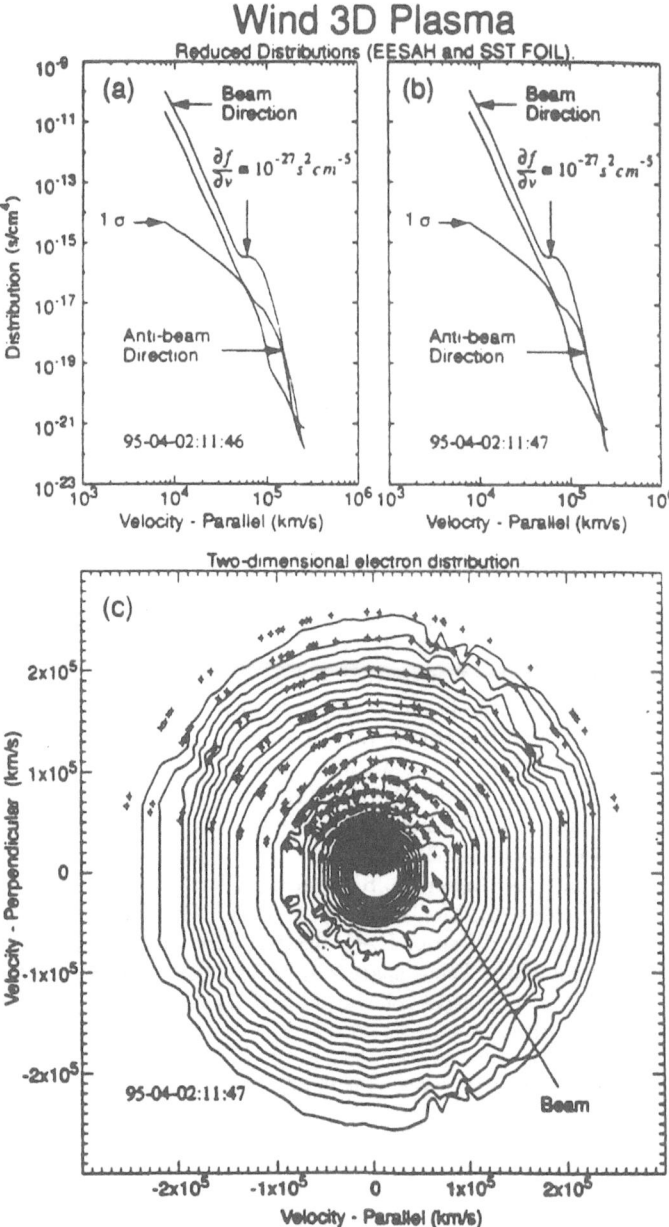

Fig. 7. (a) The reduced electron distribution in the +B direction at 11:46 on April 2, 1995 (see Fig.5). The reverse (-B) distribution and the 1σ level are also plotted. There is a small positive slope seen in both (a) and (b). Below (c) is a contour plot of the two-dimensional electron distribution. The contours represent logarithmically spaced levels. The crosses are the center velocities of the ESSAH and SST Foil detectors. The beam is at $\sim 6 \times 10^{4}$ km/s. (from Ergun et al. 1996)

in the interplanetary medium (Kennel et al., 1980; Thejappa, Wentzel, and Stone, 1995), and may play such a role here.

7 Acknowledgments

This research was supported in part by NASA grant NAG5-2815.

References

Bougeret, J.-L., R.J. Fitzenreiter, R.P. Lin, J. Fainberg, and R.G. Stone, Energetic electrons associated with solar radio storms, *Solar Radio Storms, Proc. 4th CESRA Wkshp. Solar Noise Storms*, ed. A.O. Benz and P. Zlobec, p. 320, Oss. Astr. di Trieste, Italy, 1983.

Bougeret, J.-L., M.L. Kaiser, P.J. Kellogg, R. Manning, K. Goetz, S.J. Monson, N. Monge, L. Friel, C.A. Meetre, C. Perche, L. Sitruk, and S. Hoang, *Space Sci. Rev., 71*, 231, 1995.

Dulk, G.A., and D.J. McLean, Coronal magnetic fields, *Solar Physics, 57*, 279, 1978.

Ergun, R.E., D. Larson, R.P. Lin, J.P. McFadden, C.W. Carlson, K.A. Anderson, L. Muschietti, M. McCarthy, G. Parks, H. Reme, J. M. Bosqued, C. d'Uston, T.R. Sanderson, K.P. Wenzel, M. Kaiser, R.P. Lepping, S.D. Bale, P. Kellogg, and J.-L. Bougeret, Wind spacecraft observations of solar impulsive electron events associated with solar type III radio bursts, *Astrophys. J.*, in preparation, 1996.

Feldman, W.C., J.R. Asbridge, S.J. Bame, M.D. Montgomery, and S.P. Gary, Solar wind electrons, *J. Geophys. Res., 80*, 4181, 1975.

Kennel, C.F., F.L. Scarf, F.V. Coroniti, R.W. Fredericks, D.A. Gurnett, and E.J. Smith, *Geophys. Res. Lett., 7*, 129, 1980.

Larson, D. E., R.P. Lin, K. A. Anderson, S. Ashford, C.W. Carlson, D. Curtis, R.E. Ergun, and J. McFadden, M. McCarthy, G. K. Parks, H. Reme, J.M. Bosqued, J. Coutelier, F. Cotin, C. d'Uston, K.-P. Wenzel, T.R. Sanderson, J. Henrion, J.C. Ronnet, G. Paschmann, Wind spacecraft observations of electrons from ~5 eV to ≳100 keV for an impulsive solar event and for quiet time, *Solar Wind 8 Conf. Proc.*, in press, 1996.

Lin, R.P., Nonrelativistic solar electrons, *Space Sci. Rev., 16*, 189, 1974.

Lin, R.P., D.W. Potter, D.A. Gurnett, and F.L. Scarf, *Astrophys. J., 251*, 364, 1981.

Lin, R.P., Electron beams and Langmuir turbulence in solar type III radio bursts observed in the interplanetary medium, in *Basic Plasma Processes on the Sun*, ed. E.R. Priest and V. Krishnan, p. 467, International Astronomical Union, The Netherlands, 1990.

Lin, R.P., K.A. Anderson, S. Ashford, C. Carlson, D. Curtis, R. Ergun, D. Larson, J. McFadden, M. McCarthy, G. Parks, H. R'eme, J.M. Bosqued, J. Coutelier, F. Cotin, C. D'Uston, K-P. Wenzel, T.R. Sanderson, J. Henrion, J.C. Ronnet, and G. Paschmann, A three-dimensional plasma and energetic particle investigation for the WIND spacecraft, *Space Science Reviews, 71*, 125-153, 1995.

Lin, R.P., D. Larson, J. McFadden, C.W. Carlson, R.E. Ergun, K.A. Anderson, S. Ashford, M. McCarthy, G.K. Parks, H. Reme, J. Bosqued, C. d'Uston, T.R. Sanderson, and K.-P. Wenzel, *Geophys. Res. Lett., 23*, 1211, 1996.

Maksimovic, M., V. Pierrard, and J. Lemaire, *Astron. Astrophys.*, submitted, 1996.

Pannekoek, A., Ionization in stellar atmospheres, *Bull. Astron. Inst. Neth.*, *1*, 107, 1922.

Potter, D.W., R.P. Lin and K. A. Anderson, Impulsive 2-10 keV solar electron events not associated with flares, *Astrophys. J.*, *236*, L97, 1980.

Rosseland, S., Electrical state of a star, *Monthly Notices Royal Astron. Soc.*, *84*, 720, 1924.

Saito, K., A.I. Poland, and R.H. Munro, A study of the background corona near solar minimum, *Solar Physics*, *55*, 121, 1977.

Scudder, J.D., *Astrophys. J.*, *398*, 299, 1992.

Thejappa, G., D.G. Wentzel, and R.G. Stone, *J. Geophys. Res.*, *100*, 3417, 1995.

Coronal and Interplanetary Particle Beams*

Markus J. Aschwanden[1] and Rudolf A. Treumann[2]

[1] University of Maryland, Astronomy Department, College Park, MD 20742, USA
(e-mail: markus@astro.umd.edu)
[2] Max-Planck Institut für extraterrestrische Physik, Karl-Schwarzschild-Str. 1
D-85748 Garching, Germany (e-mail: tre@mpe-garching.mpg.de) and
Dept. of Physics and Astronomy, Dartmouth College, Hanover NH 03755, USA

Abstract. In this report we attempt to synthesize the results of a series of working group discussions focused on the topics of particle beams and particle acceleration in the solar corona and in interplanetary space. We start our discussion of coronal beams with a standard flare scenario, established on recent X-ray (*Yohkoh, CGRO*) and radio observations, which constitutes a framework for the understanding of upward and downward accelerated electron beams and their secondary signatures, such as chromospheric evaporation. The second part is dedicated to interplanetary electron and ion beams, with emphasis on their relation to coronal beams, using recent spacecraft data from *Ulysses* and *Wind*. Interplanetary electron beams can often be traced back to coronal type III sources, while there is no such relation for interplanetary ion beams. In the third part, we briefly review acceleration mechanisms for coronal and interplanetary beams, separately for electrons and ions.

Résumé. Nous résumons ici une série de discussions de groupe dont le sujet était les faisceaux de particules et l'accélération de particules dans la couronne solaire et le milieu interplanétaire. Nous débutons notre discussion par les faisceaux coronaux dans le cadre d'un modèle standard d'éruption solaire, établi à la suite d'observations récentes dans le domaine des rayons X (*Yohkoh, CGRO*) et le domaine radioélectrique. Ce modèle standard est le cadre de travail nécessaire à une meilleure compréhension de la propagation des faisceaux d'électrons accélérés, ainsi que de leurs effets secondaires, comme l'évaporation chromosphérique. La seconde partie de notre synthèse traite des électrons et faisceaux d'ions dans le milieu interplanétaire, et en particulier de leur relation avec les faisceaux coronaux, grâce à l'utilisation de données récentes obtenues par les missions *Ulysses* et *Wind*. Si les faisceaux d'électrons interplanétaires sont souvent associés à des faisceaux coronaux émetteurs de sursauts de type III, ce n'est pas le cas pour les faisceaux d'ions interplanétaires. Dans la troisième partie nous résumons brièvement les mécanismes d'accélération des faisceaux coronaux et interplanétaires, en discutant séparément l'accélération des électrons et des ions.

1 Introduction

This report originated from a series of presentations and discussions about the role of particle beams (electrons, ions) as a cause of various emissions (radio,

---

* Working Group Report

hard X-rays) during solar flares, coronal mass ejections, and interplanetary disturbances. The working group discussions have been organized into three themes: the first is focused on signatures of beams in the solar corona, the second on beams in interplanetary space, and the third on the underlying acceleration processes. Due to the inhomogeneous and incoherent nature of individual contributions, this report is not designed to reflect a democratic summary, but rather attempts to synthesize the results of the discussions in a coherent "big picture" that is consistent with the latest research results in solar and magnetospheric physics.

Emphasis was put mainly on observational aspects of electron and ion beams, with theoretical considerations added only when appropriate. Thus, emission mechanisms of radio, X-ray, and γ-ray radiation are not discussed here, even if it is important for diagnostics. In particular, there is an obvious difference between beams in the chromosphere, transition region, corona, and interplanetary space, insofar as the observed radiation properties emitted by these beams strongly vary as function of the ambient plasma parameters. Theory, therefore, enters here as a very important tool to identify the exciter of the emission.

In interplanetary space, the availability of spacecraft *in situ* measurements enables one to directly detect particle beams, record their properties and even relate them to the emission processes. With the measuring technique available today it thus becomes possible to check the currently available theories of generation of radiation. It also becomes possible to infer the relevant acceleration processes at work.

2 Coronal Beams

Radio emission produced by coronal electron beams has been used as diagnostic and tracer of acceleration processes in solar flares since four decades, after Wild *et al.* (1954) established that radio type III bursts are radiation from plasma oscillations excited by discrete bunches of fast electrons, with typical exciter velocities of $v_B/c = 0.07 - 0.25$ (Dulk *et al.* 1987). Because such electron velocities require nonthermal energies in the range of 1-20 keV, the generation of coronal electron beams is inevitably associated with acceleration processes as they occur during flares. Thanks to the development and observations of broadband digital radio spectrographs (e.g. *Dwingeloo* 4-8 GHz, *Ondrejov* 1.0-4.5 GHz, *Tremsdorf* 0.04-0.8 GHz, *Zurich* 0.1-4.0 GHz), electron beams can now be traced over a large range of plasma frequencies, covering electron densities in a range of $n_e \approx 10^8 - 10^{12}$ cm^{-3}. This range of electron densities includes almost the entire corona (by mass), stretching from low-density regions in high altitudes of $\lesssim 10^5$ km down to the lower corona where active region loops ($n_e \lesssim 10^{10}$ cm^{-3}) and flare loops ($n_e \lesssim 10^{12}$ cm^{-3}) reside.

Probing this large range of coronal heights and densities with broadband observations of electron beams, we can establish a synthesized standard flare scenario, which is outlined in Section 2.1. We discuss then separately some properties of upward (Section 2.2) and downward (Section 2.3) propagating electron beams. Consequences of downward propagating particles are interactions with the chromosphere, either in form of collisional energy loss (detectable as HXR emission) or in form of heating, which can drive upflow of heated plasma into coronal flare loops. The latter process has, besides the importance as flare diagnostic in SXR, also a dramatic effect on the opacity of coronal electron beams (Section 2.4). While Sections 2.1-2.4 are entirely concerned with coronal electron beams, we devote a special section (2.5) to coronal ion beams.

2.1   A Standard Flare Scenario

The most crucial problem in the understanding of the solar flare process is the localization of the energy release and acceleration region. While it was suspected for a long time that the released energy is dissipated from the coronal magnetic field, recent SXR and HXR observations from the *Yohkoh* satellite make it more and more clear that a magnetic reconnection process is the most likely driver, occurring in an X-type or Y-type current sheet configuration in the cusp region above the SXR-bright flare loop (schematically depicted in Fig.1). While there are large uncertainties about the height, extent, and spatial fragmentation of the energy release region, it became more and more clear over the last years that the location is situated *above* the SXR-bright flare loop. Observational evidence for this scenario is provided (1) by direct imaging of heated plasma confined in the cusp region during long-duration flares (Tsuneta *et al.* 1992), (2) by imaging of (relatively weak) above-the-loop-top HXR sources in limb flares (Masuda 1994), (3) by the simultaneity of HXR emission at conjugate flare loop footpoints (Sakao 1994) together with the scaling law of electron time-of-flight distances inferred from timing measurements of HXR-producing electrons (Aschwanden *et al.* 1996b, c), (4) by observations of bi-directional (upward/downward directed) electron beams (Aschwanden *et al.* 1995), (5) plasmoid observations (Shibata *et al.* 1995) and (6) by simultaneous electron density measurements in acceleration regions and SXR flare loops (Aschwanden *et al.* 1997).

Although the detailed mechanism of particle acceleration in the reconnecting environment has not yet been identified, a variety of theoretical models have been proposed, including (1) stochastic acceleration in wave turbulence regions (e.g. Forman, Ramaty, & Zweibel 1986), (2) Fermi acceleration in MHD-turbulent cascades (Miller, LaRosa, & Moore 1996), (3) electric DC-field acceleration in reconnection outflows (Tsuneta 1995), or (4) DC-field acceleration in current sheets (Holman 1985). Most of these acceleration mechanisms are able to accelerate particles in upward and downward directions, imposed by the symmetry of the X-type topology. This produces

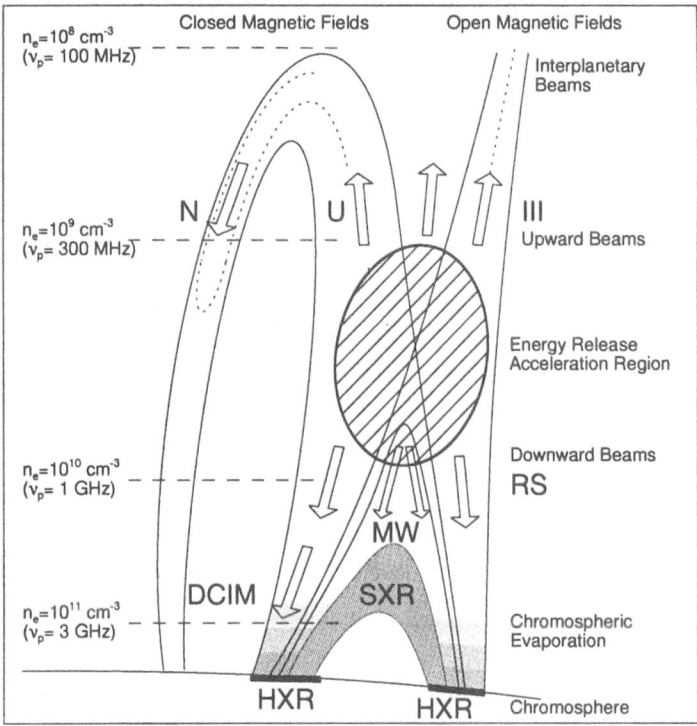

Fig. 1. A synthesized standard flare scenario: The acceleration region is located in the reconnection region above the SXR-bright flare loop, accelerating electron beams in upward direction (type III, U, N bursts) and in downward direction (type RS, DCIM bursts). Downward moving electron beams precipitate to the chromosphere (producing HXR emission and driving chromospheric evaporation), or remain transiently trapped (producing microwave [MW] emission). SXR loops become subsequently filled up, with increasing footpoint separation as the X-point rises.

upward directed electron beams, either propagating along open field lines (metric type III bursts) or closed field lines (metric type U bursts), which occasionally are observed to be mirroring (metric type N bursts). The downward accelerated electrons can form reverse-drifting decimetric type III bursts, but are rarer observed than metric type III bursts, probably because of the higher free-free opacity that suppresses plasma emission in the dense lower corona. However, some reverse-drift bursts have been observed up to 8 GHz (Benz *et al.* 1992).

Analysis of the timing of HXR footpoint emission has revealed that the nonthermal HXR emission consists of two components (Aschwanden *et* al. 1996a, b): (1) a pulsed ($\lesssim 1$ s) HXR component produced by electrons that precipitate directly from the acceleration site to the chromospheric footpoints (according to energy-dependent delays that correspond to time-of-flight differences), and (2) a smooth HXR component produced by temporarily ($\approx 1-10$ s) trapped electrons (according to energy-dependent time delays that correspond to collisional deflection time differences). The temporarily trapped electrons that mirror along closed field lines are also responsible for the microwave emission, which consequently peaks a few seconds later than the nonthermal HXRs from the footpoints.

The precipitating electrons and ions heat the chromospheric footpoints and drive the so-called "chromospheric evaporation" process at the footpoints (Doschek *et al.* 1980). The heated plasma, driven by the local overpressure, flows upward into the flare loop with velocities of $\approx 100-300$ km, filling it up on time scales of $\approx 20-30$ s. The scenario depicted in Fig.1 should be understood as a single snapshot during a dynamic evolution. Magnetic reconnection is expected to proceed in upward direction (Kopp & Pneumann 1976), so that the instantaneous X-point is increasing in height and the magnetically connected footpoints become more separated. During this continuing evolution, closed flare loops fill up subsequently by chromospheric evaporation, with increasing footpoint separation. At some point, reconnection stops abruptly in impulsive flares and the cusp-shaped flare loops relax into a dipolar-like structure. Only in long-duration flares, chromospheric evaporation is sustained sufficiently long to fill up the cusp region up to the reconnection point, so that it can be clearly seen in SXR (Tsuneta *et al.* 1992).

2.2   Upward Propagating Electron Beams

Normal-drifting type III bursts (indicating upward propagation because of the decreasing electron density) are generally found at metric frequencies ($\nu \lesssim 300$ MHz), while reverse-drifting type III bursts (indicating downward propagation because of the increasing electron density) are dominantly observed at decimetric frequencies ($\nu \gtrsim 300$ MHz). In some flares, a clear separatrix frequency can be observed between normal- and reverse-drifting bursts, e.g. at 620 and 750 MHz for two flares (Aschwanden, Benz, & Schwartz 1993). A straightforward interpretation is that such a separatrix frequency corresponds to the plasma frequency in the acceleration region, which is typically found to correspond to electron densities of $n_e \approx 10^9 - 10^{10}$ cm^{-3} (see Fig.1). Since electron densities in the SXR-bright flare loops are observed to be typically an order of magnitude higher ($n_e \approx 10^{11}$ cm^{-3}), the acceleration regions are most likely located about two scale heights *above* the SXR flare loop.

Although our standard flare scenario implies some symmetry in upward and downward direction, the acceleration efficiency or number of accelerated electrons is not necessarily symmetric in both directions. Lin (1974) reported that the electron ratio is estimated to be of order $n_{up}/n_{down} \approx 10^{-2} - 10^{-3}$, comparing the measured number of electrons in interplanetary type III bursts with the required electron number to explain the coronal HXR emission. In our discussion we did not came to a solution of the problem, to decide whether the asymmetry is associated with the acceleration mechanism or with the magnetic geometry. It was also argued that closed field lines could redirect a large number of upward escaping electrons back to the lower corona, as it is observed in type U bursts. Karlicky, Mann, & Aurass (1996) presented new observations of type U and N bursts. In numerical models they demonstrated that electrons could be reflected not only by magnetic mirroring, but also by whistler turbulence. Equally, the same mechanism could be applied to reflect upwardly escaping electrons on open magnetic field lines, which could explain part of the discrepancy in the upward/downward ratio. Anisotropic electron velocity distributions that contain a forward beam and a weaker backward beam were also observed in interplanetary space (Lin 1996, Biesecker 1996), indicating backscattering. Another explanation for the asymmetry of upward/downward accelerated electrons could be the trapping efficiency of the magnetic geometry in the acceleration region: The dominantly closed field lines below the X-type reconnection point may provide a longer confinement of the accelerated particles than the preferentially open field lines above the X-type reconnection point. None of these questions could be conclusively answered in the discussions of the working group.

The standard flare scenario depicted in Fig.1 can be modified in various ways. Different geometries can be obtained by inclining the external magnetic field by some angle to the vertical. Such a unification scheme was proposed by Yokoyama & Shibata (1996), to explain SXR jet events in the same reconnection framework as cusp-shaped flares. In such inclined geometries, the upflowing plasma driven by chromospheric evaporation can also have access to inclined or horizontally oriented open field lines. Kundu *et al.* (1995a, b) reported about metric type III bursts that were found to be aligned with collimated SXR jets, both being channeled by a common bundle of open magnetic field lines.

2.3   Downward Propagating Electron Beams

Downward propagating electron beams can be traced either by reverse-drifting decimetric radio bursts or by correlated HXR pulses. Correlations between these two signatures in radio and HXR have been studied with exhaustive statistics (Aschwanden *et al.* 1995 and references therein) and it was found that the starting time of reverse-drift bursts coincides with HXR pulses of equal duration ($\lesssim 1$ s) within $\lesssim 0.1$ s in many cases.

While radio emission from coronal electron beams is associated with electron velocities of $v_B/c = 0.07 - 0.25$, or kinetic energies of 1-20 keV, respectively, nonthermal signatures in HXR were chiefly detected at higher energies, say $\gtrsim 25$ keV. However, Lin and Feffer identified recently for the first time low-energy (7-25 keV) HXR spikes from the *BATSE* spectroscopic detectors. Similar observations at $\approx$ 10 keV have also been presented by Norma Crosby from *WATCH/GRANAT* at this meeting. These rapid HXR spikes are most likely nonthermal signatures, attributable to downward propagating electron beams. The collisional energy loss time for such low-energy electrons limits the lifetime of such beams to $\lesssim$ 100 ms in flare loops with densities of $n_e \approx 10^{11}$ cm^{-3}.

A novel diagnostic of downward propagating electron beams is the measurement of energy-dependent time delays of 25 − 250 keV HXR photons. These HXR photons are produced by 50 − 500 keV electrons in the thick-target model, which have relativistic velocities of $\beta = v/c = 0.4 - 0.9$. The time-of-flight difference of these electrons propagating between the coronal acceleration site and the chromospheric energy loss site amounts to 20-200 ms in typical flare loops and can be measured with a high-precision cross-correlation technique between different energy channels from the most sensitive *BATSE/CGRO* detectors. Employing this technique, Aschwanden *et al.* (1996c) discovered a scaling law of $l'/s = 1.4 \pm 0.3$ between the (projected) electron time-of-flight distance l' and the flare loop half length s. If a symmetric situation is assumed for magnetic field lines from the acceleration site to conjugate flare loop footpoints (as indicated by Sakao's [1994] simultaneity measurements), the required height of the acceleration site is $h/s = 1.6 \pm 0.4$. This observation strongly supports the scenario with an acceleration site *above* the SXR-bright flare loop (Fig.1). This measurement technique allows us for the first time to pinpoint the starting point of downward propagating electron beams. In the case of a fragmented acceleration site (Benz 1985; Vlahos 1993; Bastian & Vlahos 1996), the measured position would correspond to the geometric centroid of all elementary acceleration sites.

2.4   Chromospheric Evaporation

Precipitating electron beams lose their energy in the chromosphere, where most of the energy is deposited into heating of the local chromospheric plasma, while only about a fraction of $\approx 10^{-5}$ goes into HXR bremsstrahlung photons. This intense heating drives upflow of chromospheric plasma, which has a typical electron density of $n_e = 10^{10} - 10^{11}$ cm^{-3} and an electron temperature of $T_e \approx 15$ MK (Antonucci *et al.* 1984), as measured from blueshifted Ca XIX lines. This secondary process is mainly responsible for the brightest signature of a flare in SXR, in form of a coronal loop filled with a 15 MK plasma. Thermal bremsstrahlung from this heated plasma contributes also to a smoothly-varying component of the microwave and millimeter emission, commonly peaking during the decay phase of the flare. Valery Zaitsev presented model calculations, comparing the relative contributions of gyrosynchrotron emission and bremsstrahlung at microwave/millimeter wavelengths, and found that bremsstrahlung dominates the spectrum at higher frequencies (e.g. $\gtrsim$ 40 GHz for typical parameters). Similar model calculations of microwave spectra including both the traditional gyrosynchrotron radiation as well as the usually neglected Coulomb bremsstrahlung of the thermal flare plasma were also presented by Joachim Hildebrandt (Hildebrandt *et al.* 1997). Numerical fits of bremsstrahlung emission to the gradual component of observed microwave time profiles were also shown by Silja Pohjolainen during this meeting. However, the impulsive peaks observed at millimeter wavelengths correlate well with the HXR and γ-ray time profiles, and are, therefore, produced by the directly precipitating energetic particles, as emphasized by Kundu (Kundu *et al.* 1994).

Chromospheric evaporation has also a dramatic influence on the opacity of precipitating electron beams. When downward propagating electrons intercept with the upflowing hot plasma, the free-free opacity ($\tau_{ff} \propto n_e^2 T^{-3/2} \nu^{-2}$) changes dramatically. The front of the upflowing plasma becomes transparent to plasma emission due to the temperature increase, while the bulk flow behind the front becomes opaque tue to the density increase. This effect was discovered in decimetric radio bursts with slow-drifting high frequency cutoffs that match the velocity of the upflowing chromospheric plasma (Aschwanden & Benz 1995). Such decimetric bursts (indicated with DCIM in Fig.1) are also expected to have an almost infinite frequency-time drift rate because of the very short density scale height across the evaporation front. Because of these very specific characteristics in the drift rate and high-frequency cutoff, these DCIM bursts provide a sensitive tracer and diagnostic of the chromospheric evaporation process (see also Bentley 1996).

Based on these studies and our standard flare scenario (Fig.1), we have to distinguish between three different populations of electrons to understand and model radio emission in flares: (1) directly precipitating high-energy electrons which produce fast (sub-second) HXR, γ-ray, millimeter, and microwave pulses, (2) temporarily (1-10 s) trapped high-energy electrons which are responsible for a delayed, smoothly-varying HXR emission and gyrosynchrotron emission in microwaves, and (3) thermal electrons evaporated from the chromosphere and subsequently trapped in the flare loop, which produce SXR emission and thermal bremsstrahlung at millimeter wavelengths.

2.5   Coronal Ion Beams

Indirect evidence for the existence of coronal ion beams is provided by the occasional emission of γ-ray lines during solar flares and by the concomitant (very rare) ground-based detection of prompt neutrons (Forrest and Chupp 1983; Nakajima *et al.* 1983). Such emission requires nuclear interactions in the deep layers of the solar atmosphere, like neutron capturing, pion production, and other decay processes, which all require energetic ions with energies up to 1 GeV. These ions cannot exist a priori in the solar atmosphere (because no continuous line emission is observed), but must have been produced during the initial or a later phase of the flare by some unknown acceleration mechanism that speed up ions from thermal to relativistic energies. In addition, these ions should be collimated to penetrate into chromospheric depths, where densities are suitably high for nuclear processes. Thus, the mere existence of line emission is sufficient evidence for the existence of very energetic ions under flare conditions. The most reasonable assumption is that these ions are accelerated in the corona at the flare explosion site, being bundled and streaming down to the chromospheric γ-ray emission site. Alternatively, a local fast heating or acceleration mechanism in the chromosphere would be required, which is difficult to realize in a highly collisional plasma.

Simnett (see also Simnett 1994, 1995) reported that 40 keV$-$40 MeV photons are synchronized within ± 1 s during γ-ray line flares (Mandzhavidze & Ramaty, 1992), which requires that GeV protons must have been produced on a very fast time scale in order to permit for pion production and decay. On the other side, extended γ-ray emission was observed up to 8 hours, which requires relativistic long-duration trapping of ions. The energetic importance of ions is also indicated by the high proton-to-electron ratio of $> 10^2$, found at γ-ray energies above 30 MeV.

Most of the energy is, however, stored in lower energy protons, at energies of $0.1-1$ MeV. Lyα and polarization measurements of the $H\alpha$-line may give indirect information about the abundance of these protons during chromospheric flares. New indications that a large fraction of the flare energy is contained in accelerated ions comes also from Ne/O abundance ratios measured in γ-ray lines (Ramaty et al. 1995). Concerning the beaming of such ions, little information is obtained. Slowly-drifting type III-like radio bursts (Benz & Simnett 1986) as well as type I burst emission (slow drift, fundamental) have been brought into context with ion beams, but no convincing evidence exists so far that radio emission is produced by ion beams. Electrons are probably more suitable to produce the observed high-frequency emission, because ion growth rates are too low at these high frequencies.

In summary, coronal ion beams are very likely to exist and may carry most of the energy and momentum during solar flares. They certainly are responsible for production of nuclear γ-ray lines, but they seem not to produce signatures in HXR or radio wavelengths.

3   Interplanetary Beams

3.1   Electron Beams

Electron streams in interplanetary space are defined as streams which propagate on the background of the solar wind flow, which itself is a particle stream of varying density, temperature, and velocity ($\approx 300 - 1200$ km/s). Since *Ulysses'* measurements, high-speed flows are known to be connected with coronal holes through open field lines, either at high solar latitudes or via interequatorial crossings. Low-speed streams originate from equatorial regions between $\pm 30°$ solar latitude.

Electron beams originating in the solar corona propagate in anti-sunward direction along the interplanetary magnetic field. Sometimes part of those beams may be scattered in backward direction. Electron beams can also be generated in interplanetary space. Dominant sources are planetary bow shock waves, at Earth, Jupiter, or in interplanetary shocks. Electrons of solar wind origin are reflected from these shocks and flow generally upstream into solar direction, against the solar wind along the interplanetary magnetic field lines. For example, Jupiter bow shock-reflected electron beams are known to propagate upstream up to the Earth.

Observations. Interplanetary electrons have been detected with the *Wind* spacecraft (Lin 1996) over the entire energy range from ≈10 eV to ≈100 keV during solar minimum. Three different populations of the electron spectrum have been identified: (1) a core solar wind thermal component at low energies with an approximate Maxwellian distribution, (2) a halo distribution extending into the keV range, and (3) a further high energy super-halo in the few keV to ≈100 keV range (similar as during flare conditions, but at a lower flux level).

This background distribution is obscured during the passage of transient electron beams impulsively accelerated at the Sun and escaping into the interplanetary medium, lasting at the *Wind* spacecraft for tens of minutes up to hours. These beams have typical energy maxima around 1 keV, and their reduced parallel electron distribution function, constructed with time resolution of 96 s, exhibits the usual core/halo distribution, and sometimes a slightly positive slope in anti-sunward direction. The relatively long integration time of 96 s may obscure steeper positive slopes on shorter time intervals.

Tracing those beams from the *in situ* position at the *Wind* spacecraft back to the solar corona, seems to indicate an acceleration site relatively high in the upper corona (Benz & Lin 1996). The localization of the acceleration site bears uncertainties arising from fluctuations in the magnetic field and deviations from the Parker spiral, as measured by *Ulysses*. An acceleration site in the upper corona is not in contradiction with our standard flare scenario (Fig.1), where magnetic reconnection X-points can form far above the SXR-bright flare loops, which are located in the lower corona. However, as discussed below, indirect observations suggest that in many cases the acceleration site of interplanetary type III electron beams may be deeper in the corona.

The energy range of electrons (≈ 0.5 to $\gtrsim 100$ keV) measured in interplanetary beams is consistent with those of solar (e.g. metric) type III bursts (typically below 10 keV). While higher-energy electrons are required for solar flare X- and γ-ray emission, there are also high-energy electrons present in the halo distribution function in the interplanetary background population, probably produced by flare-independent acceleration mechanisms. It has been shown (Buttighoffer *et al.* 1995) that the beams propagate in well defined plasma structures, well isolated from the surrounding plasma and

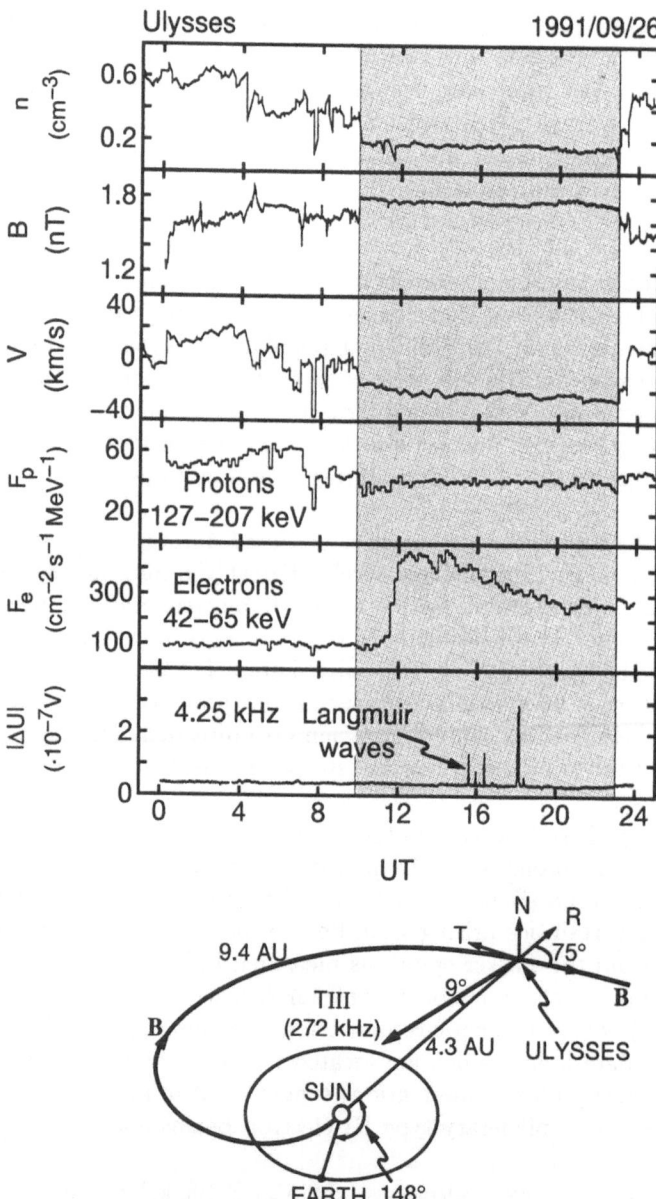

Fig. 2. *Ulysses* particle and field data inside a quiet low density stream with high energetic electron fluxes at 4.3 AU (courtesy A. Buttighoffer and M. Pick). The shaded region indicates (the propagating channel of) the quiet plasma stream. The lower part of the Figure shows the position of the spacecraft and the path of the electron stream along the Parker spiral.

bearing quiet magnetic fields (Fig.2). In this particular case the beam propagated practically scatter-free along the spiral magnetic field over a length of 10 AU indicating that it was extraordinarily stable. The stabilization time of a structure of 10^7 km length is about 140 hours, given an estimated velocity difference of 20 km/s.

Type III solar electron beam fluxes are non-stationary by nature. At constant energy, these fluxes exhibit fluctuations on time scales of tens of seconds up to minutes which may reflect either fluctuations in the source injection mechanism or local modulation. In the first case, they map the time variation of the magnetic reconnection dynamics, while in the second case, they are caused by firehose-like instabilities driven by the heat flux.

There is plenty of indirect observations of shock-reflected beams near the Earth's and Jupiter's bow shocks, provided by spacecraft like *ISEE*, *Ampte/IRM*, *Galileo*, and others. The information is obtained from tracing of the magnetic field tangential to the shock wave and from beam-generated Langmuir waves. Most shocks are not strong enough to produce a sufficient number of reflected electrons to be detectable in the distribution function.

Relation to Interplanetary Type III Bursts. The relation between interplanetary electron beams and locally excited (interplanetary) type III radio bursts has been probed with *in situ* spacecraft, e.g. *ISEE, Galileo, Ulysses*, and *Wind*. Langmuir wave emission of highly fluctuating amplitude has been observed in connection with positive average slopes of the anti-sunward electron beams. Harmonic radio wave emission has been recorded in some cases together with radiation at the fundamental. These observations support the view that large-amplitude, but fluctuating, Langmuir waves excited by the bump-in-tail instability of interplanetary electron beams coalesce to generate harmonic interplanetary type III radiation thus confirming the general wave-wave emission theory.

The identification of the type III burst radiation mechanism is still hampered by insufficient temporal, spatial, and frequency resolution of plasma waves. Observationally, wave coalescence during type IIIs proceeds in the presence of large-amplitude bursts of ion-acoustic waves. But the correlation between the bursty emission in both Langmuir and ion-acoustic waves is poor. Thus it has not yet been clarified whether the ion-acoustic bursts participate in the radiation process, constituting a necessary requirement of the radiation mechanisms, or are simply a by-product caused by the heat flux transported in the interplanetary electron streams.

In weakly turbulent type III radiation via merging of Langmuir waves, counterstreaming Langmuir waves are required. Since the beam excites waves only in one direction, scattering of Langmuir waves is needed. This could be achieved either by nonlinear scattering in Langmuir wave collapse or by scattering off background ions or ion acoustic waves. It thus seems that ion-acoustic waves are an ingredient of the radiation process.

Fundamental emission has been reported by Reiner *et al.* (1995) from *Ulysses* observations. In a statistical study (Leblanc *et al.* 1996) of the lower cut-off frequencies of interplanetary type III bursts in the solar wind, it was found that this cut-off frequency never falls below $f_0 \approx 9$ kHz, while at higher frequencies closer to the Sun it coincides with the plasma frequency $f_0 \geq f_{pe}$. Given that the lowest frequency at which type III radiation can be excited is f_{pe}, one concludes that interplanetary type III bursts may be generated at the fundamental for distances close to the Sun. The reason for the mysterious 9 kHz lower cut-off is not clear. The simplest explanation is that, at a distance corresponding to 9 kHz, the type III electron beam ceases to maintain a sufficiently steep positive slope to overcome Landau damping. This should illuminate the process of beam stabilization.

Construction of the density profile in the solar wind from the observed type III spectrum has confirmed the r^{-2} variation in solar wind density. However, scattering of radiation has been observed sometimes at large distances from the Sun, indicating the presence of very large loops or magnetic inhomogeneities in the solar wind by *Ulysses* and *Ampte/IRM* (Treumann & LaBelle 1996). Such cases seem to be related to coronal mass ejections (e.g., Gosling *et al.* 1987).

Relation to Coronal Type III Bursts. Naively, one believes that interplanetary electron beams are but the prolongation of solar type III electron beams into interplanetary space. The connection between both interplanetary electron beams and type III bursts and coronal electron beams and solar type III bursts is not that simple. Benz & Lin (1996) have undertaken an investigation of this connection and have concluded that a coronal source of interplanetary electron streams related to solar type III bursts exists. The most sophisticated analysis of this relation is provided by Poquérusse *et al.* (1996), who used the newly-developed technique of the *Nançay Artemis* spectrograph to correlate ground-based observations of solar type III radio bursts with *Ulysses* measurements of interplanetary type III radiation. Poquérusse *et al.* (1996) find that to almost every interplanetary type III burst, whose spiral field line is connected to the visible disk or limb of the Sun, belongs a *group* of coronal type III bursts.

Figure 3 shows an example of such an *Artemis–Ulysses* reconstruction of the connection between solar and interplanetary type III bursts. Extending the slope of the leading edge of the interplanetary type III emission across the measurement gap down to the higher frequencies of the coronal band, the origin of the interplanetary type III burst is found to coincide with the end of the coronal type III group.

This work must be considered as a major scientific achievement. It has a number of very important implications mentioned by Poquérusse. The first is that almost all interplanetary type III bursts are rooted in coronal type III bursts. Their electron beams originate in the solar atmosphere, propa-

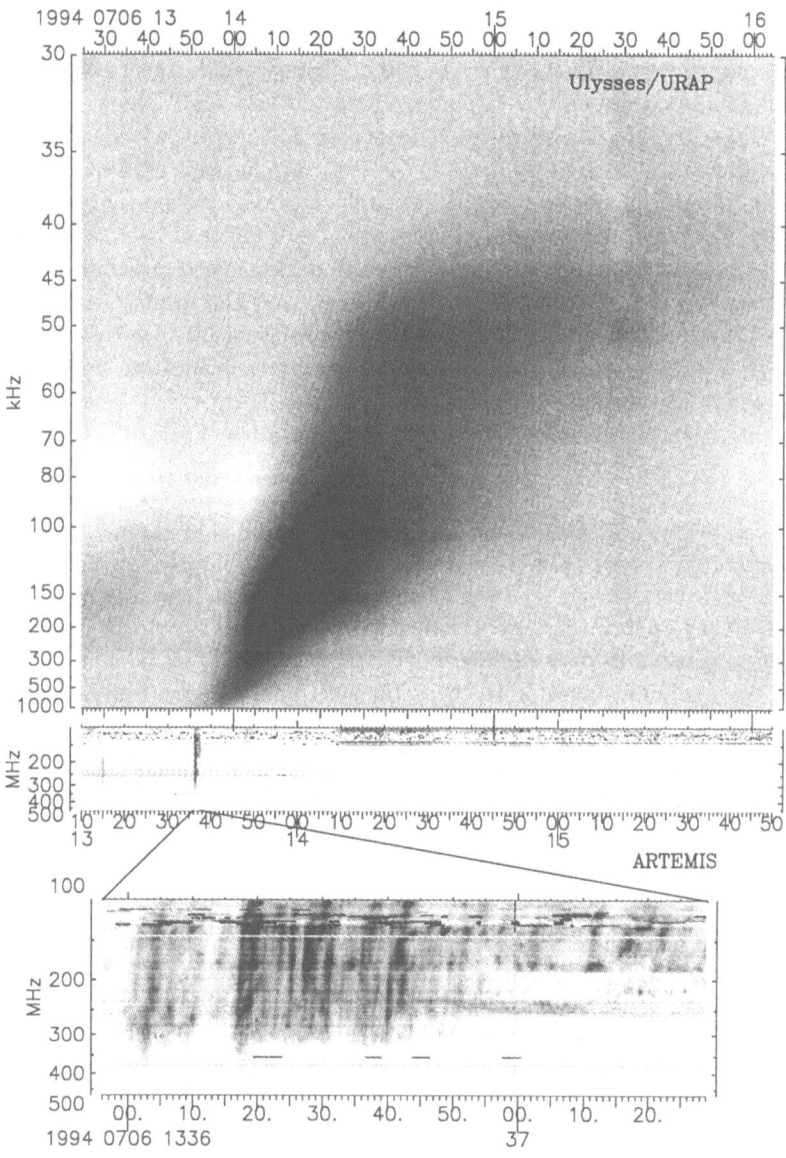

Fig. 3. *Artemis- Ulysses* combination of a group of solar metric type III bursts and an interplanetary type III radio burst. The latter is represented with frequency increasing down as is common for solar type III bursts. The scale is chosen in such a way that its prolongation to higher frequencies gives the starting point of the interplanetary burst in the corona. It is seen that this point coincides with the low-frequency starting point of the solar type III group proving that the interplanetary burst is nothing but the extension of this group (courtesy of M. Poquérusse). The lowest panel shows the high time resolution of the type III group identifying it as a large group consisting of many members.

gate stably all the way across the corona. Inversely, not every solar type III burst, even if strong, produces an interplanetary type III burst. To become an interplanetary electron beam requires additional conditions to be satisfied, presumably open magnetic fields and possibly other conditions as well.

Type III bursts giving rise to interplanetary type III bursts and electron beams usually appear in groups. These groups are not resolved at low frequencies in interplanetary space. In the solar atmosphere, each of the bursts lasts up to several seconds only. These groups of electrons obviously merge into one broad electron beam due to velocity dispersion, when entering interplanetary space. It should be noted, however, that the minute-scale oscillations of the electron flux in interplanetary type III electron beams reported above may still memorize the origin of the beam, resulting from many bursty injections in the corona.

Stereoscopic investigations of the emission sites of solar type III bursts related to interplanetary observations (Maia *et al.* 1996, Buttighoffer *et al.* 1996, Pick *et al.* 1995 a, b, Manoharan *et al.* 1996) indicate injection sites widely separated in latitude. It has even been found that the magnetic configuration may belong to a loop which connects the southern to the northern hemisphere over a large latitudinal distance across the equator. The implication is that open interplanetary flux tubes are built up out of small open flux tubes rooting in many remotely separated coronal locations.

As a counter-example to the solar-interplanetary relation, a case was mentioned, where no relation to any solar activity at radio frequencies could be found, during a major storm of interplanetary type III bursts on May 15, 1985, lasting for more than 8 hours (LaBelle & Treumann 1996), recorded by *Ampte/IRM* at frequencies below 5.6 MHz. The magnetic field was possibly connected to the backside of the Sun at this time.

Relation to Shock Waves. Electron beams are well known to be connected with observations of shock waves in interplanetary space. They are caused by reflection of solar wind electrons from the point where the interplanetary magnetic field is tangential to the curved shock front. Their manifestation is in local Langmuir wave emission. High time and frequency resolution observations performed *in situ* with *ISEE* and *Ampte/IRM* in front of the Earth's bow shock and with *Voyager* close to Jupiter's bow shock, as well as in the vicinity of interplanetary shocks with *Ampte/IRM* have shown that Langmuir wave emission is strongest along the tangential magnetic field line where the beam is fast and about monoenergetic. Deeper inside the foreshock region, the beam becomes diffuse, ring-like, and the emission is weaker.

The Langmuir waves along the tangential field line are bursty in nature and broadband, extending to frequencies below and above the local plasma frequency, both being thermal effects. Bursty ion-acoustic waves are also found in these cases. There is a correlation between their appearance and that of Langmuir bursts, but this correlation is not convincing enough to unambiguously support the collapse of Langmuir waves. Part of the ion-acoustic waves may be generated in a different way, or burnt-out cavitons may be detected sometimes.

In many cases, locally generated harmonic radiation has been detected in connection with bursty Langmuir waves. Splitting of this emission into two closely-spaced bands is occasionally found (Treumann et al. 1986). Fundamental radiation is often obscured by the locally detected Langmuir waves, from which it can be distingusihed only by non-available polarization measurements. The most interesting observation is that in all cases no radiation has ever been detected to come from the shock front or from the compressed region behind the shock. Langmuir emission behind the shock is at a thermal level, because electron beams are absent there. In the shock front itself, there is strong low-frequency turbulence at ion-acoustic frequencies. The electron distribution shows a weak beam in the shock ramp, which may give rise to electron acoustic waves with frequencies between $f_{pe}/2$ and f_{pe}, which in turn may cause fundamental radiation when merging. But the intensities of this radiation seem to be low.

3.2   Ion Beams

Interplanetary ion beams behave collision free. The only radiation to which they contribute is plasma emission in plasma modes that cannot escape. These modes may participate in generation of escaping radiation at the fundamental in wave-wave interactions. But the advantage of interplanetary ion beams over coronal ion beams is that they can be measured in situ.

Solar Wind Ion Observations. From in situ measurements of solar ions escaping into the solar wind one can infer about the presence and acceleration of energetic ions in the solar atmosphere. The solar wind ions, mainly the proton component of the solar wind, have energies of order of a few keV bulk energy, but at low temperature. Energetic ions (protons) are emitted from the Sun during solar flares. The energetic component superimposed on this stream covers the energy range up to solar cosmic ray energies of > 20 GeV protons.

Ulysses detected high energetic ion fluxes during the passage of the spacecraft across the ecliptic (Simnett 1986, 1994, 1995), which seem to persist during the entire passage of the satellite. These fluxes fluctuated strongly, but there was a persistently enhanced energetic ion flux background, which, if believable, indicates that even in the ecliptic there seems to exist an energetic component of ion fluxes which resembles that of the Wind measurements of electron spectra. If this is true, the inner heliosphere acts as a reservoir of energetic protons which can trap protons over very long times, of the order of one month. Pick pointed out that the short period when Ulysses swang through the ecliptic coincided with a period of very strong solar activity, when energetic ions are emitted from the Sun in large number. *Ulysses* observations thus suggest that presence of energetic ions in interplanetary space is mainly warranted during favorable connectivity with solar flare regions (Roelof et al. 1992).

Protons of 1 MeV have been detected, and > 45 MeV protons have been measured at solar maximum. The lower energy part may be due to acceleration at interplanetary shocks. In order to identify interplanetary protons as of solar origin, their spectral shape must agree reasonably well with those spectra that are expected for solar flares. Spectra flatter than power laws of index -2 are expected for the low-energy part. Such power-laws have been observed below 250 keV. *Ulysses* detected a spectral peak at 270 keV for interplanetary protons, with a pitch-angle distribution suggesting coronal injection. The cut-off of the spectrum was at a few MeV. Hence, sub-MeV protons are energetically dominant. The proton-to-electron ratio at a fixed energy is an indication for the origin of protons. This ratio was found to be of order $< 3 \times 10^2$, in rough agreement with the expectations for γ-flares.

Shock Evidence. Planetary bow shocks provide the best evidence for interplanetary proton beams. These beams have been measured to escape from the shock (Thomsen, 1985). Their energy in the solar wind frame is four times that of the solar wind, which is a few keV. But due to different mechanisms of acceleration, these beams are readily scattered into kidney beam distributions and finally into halos around the solar wind. The available energy in the planetary bow shocks is only of the order of the solar wind energy, and the final energy which can be picked up by protons in the acceleration process is not very high. If protons are scattered many times, i.e., if they survive sufficiently long in front of the shock, they may be accelerated to energies of order of hundreds of keV. The problem is to keep them long enough in front of the shock and to scatter them frequently.

Ion beam measurements are familiar from observations of particle fluxes in the auroral zone of the Earth where the ions are accelerated by field-aligned potential drops and wave fields. Both mechanisms yield auroral energies and anisotropic pitch-angle distributions.

Wave Signatures. The wave signature of ion beams in interplanetary space is best studied in front of planetary bow shocks and in the shock reformation process itself. It is found that shock-reflected ion beams mixing into the foreshock excite magneto-acoustic and Alfvén waves (Thomsen *et al.* 1990), either resonantly or non-resonantly. The highest growth rate is found for the left-polarized mode, which may reverse its polarization in the streaming plasma. This mode has a compressible component and thus propagates in the fast mode. It evolves to large nonlinear amplitudes and may form shocklets and solitary waves, which serve as the basic scatterer of the ion component. The frequency of these waves is below the ion cyclotron frequency, but their spectrum is a power law spectrum. The waves play an important role in ion isotropization and are also responsible for shock reformation. Whether or not they contribute to radiation processes is not known.

Relation to Coronal Phenomena: Flares and CME's. Pick (Pick *et al.* 1995a, b) reported about observations when *Ulysses* passed a Coronal Mass Ejection event (CME) in interplanetary space which was *not a detached magnetic cloud*. During this time period, ions streaming from the Sun into the CME had been detected. This observation indicates that sometimes very large flux tubes build up extending from the Sun out to the CME material and that particle ejection from the Sun into those tubes must persist over long times.

4   Acceleration Mechanisms

4.1   Acceleration in the Corona

The main mechanisms of acceleration in the corona is believed to be related to dissipation of free energy stored in active magnetic field configurations of temporarily complex geometry (e.g. Brueckner and Bartoe 1983, Dere *et al.* 1991, Tsuneta *et al.* 1992). Energy is transformed partially into plasma heating and partly into bulk plasma acceleration. During reconnection, bulk acceleration by the inductive magnetic field accelerates both electrons and ions to about the Alfvén speed, v_A. Assuming plasma densities of the order of $10^{11} > n > 10^9 \text{cm}^{-3}$ and annihilation of, say, a $B \approx 10$ Gauss magnetic field implies velocities between $70 < v_A < 700 \text{ km/s}$. This corresponds to bulk proton energies between $15 \text{ eV} < m_i v_A^2/2 < 1.5 \text{ keV}$, far below the required and observed ion energies in solar flares. This simple estimate shows that bulk energization of plasma in reconnection cannot produce the high energies inferred from X- and γ-ray observations. But velocities in this range are in good agreement with line broadening observed in optical emission, known since *Skylab*. Beams are thus no good indication or confirmation for the existence of reconnection, while optical line profile asymmetries are. Explaining the existence of beams requires more sophisticated theories, i.e. selective particle acceleration. In reconnection events in the Earth's geotail particle beams of medium energy along the separatrix have been detected. Kinetic simulations confirm that the inductive reconnection electric field may accelerate a very small fraction of particles to high energies. A review of some flare acceleration mechanisms is found in Chupp (1996).

Electron Acceleration. It is relatively easy to accelerate electrons parallel to the magnetic field in the two cases when field-aligned electric potentials can be maintained for sufficiently long time or in resonant interaction with waves. Both mechanisms encounter problems. It is difficult to maintain large potential drops of the order required for accelerating electrons up to several 100 keV or MeV energies along the magnetic field. In plasmas of background thermal energy of tens to hundreds of eV it is unreasonable to encounter total potential drops larger than thermal (several 100 V), because they correspond to total depletion of the density, $\Delta n/n \approx e\Delta\phi/k_B T_e \sim 1$, unless they are forced by some violent external mechanism. Realistic potential drops will be of the order of a fraction of the thermal drop over distances of several Debye lengths.

At coronal altitudes, this distance corresponds to < 10 cm. For potential drops of $\Delta\phi \approx 0.1$ V over this distance one requires more than 10^6 of such structures of same polarity in a chain along a field line. Assuming a spacing of only 10 cm between the structures yields a total length for the total potential drop of > 1000 km in the collisionless plasma. This length is not unreasonable. Hence, microscopic double layers may well exist and add up their potential drops to an overall potential high enough to accelerate electrons to the required energies found in HXR and γ-ray flares. Double layers are presumably ion holes or deformed non-symmetric ion-acoustic as well as lower-hybrid solitons, which may develop by nonlinear interaction in plasmas. Low-frequency waves like kinetic Alfvén waves may also form similar structures, but their length-scales are too large.

Another possibility of generating field-aligned potential drops is via formation of a resistive conduction front (Brown *et al.* 1979, Batchelor *et al.* 1985) which keeps a thermal plasma confined and moves at about the ion sound velocity along the loop. Across the conduction front, electric fields are maintained by anomalous collisions ν_a, yielding an anomalously high resistivity $\eta = m_e \nu_a / e^2 n_0$ and giving rise to an electric field $E = \eta j_\parallel$ in presence of a field-aligned current $j_\parallel$. Accelerated electrons are then produced as runaway electrons. Batchelor *et al.* (1985) have shown that such acceleration may account for the electron component in flares.

Stochastic ("second-order Fermi") acceleration of electrons is attributed to interaction with waves. For an energetic electron propagating parallel to the strong external magnetic field, the condition for resonant interaction is that the Doppler-shifted phase velocity of the wave matches the electron velocity. In general this condition can be satisfied for near-perpendicularly propagating electrostatic waves like lower-hybrid and ion-cyclotron waves. The latter wave is not efficient in accelerating electrons because of its narrow spectrum. Lower-hybrid waves having broad spectra are well-suited for particle acceleration. Wu *et al.* (1981) have developed a theory for such stochastic acceleration of electrons in extended lower-hybrid wave fields under non-relativistic conditions. The relativistic extension has not yet been presented. An originally Maxwellian distribution of electrons slowly evolves towards an energetic tail. Strong wave fields are required and long times needed for reaching high energies. These requirements and the infinite extension of the wave field are unrealistic.

Dubouloz *et al.* (1995) have presented a model where the wave field is localized due to nonlinear self-interaction and evolution into cavitons. A single caviton in this case accelerates electrons into beams during the transition time of the electrons across the caviton. The energy reached with one single caviton is of the order of 16 times the thermal energy. Core electrons are not accelerated by this mechanism. Numerical simulations for lower-hybrid fields have been performed by Retterer *et al.* (1995). An extension of the theory to many cavitons is required and may lead to selective effects of accelerating electrons. The case of many cavitons is more realistic because an initial wave disturbance will readily evolve into many cavitons which are distributed with respect to their amplitudes and speeds. This produces a caviton plasma which has a much harder spectrum than the initial wave field.

A recent model by Miller *et al.* (1996) uses extended cascading of fast waves to demonstrate that electrons can also be accelerated in low-frequency waves. For sufficiently long wavelengths (small wavenumber) of these waves, the resonance condition can be satisfied as well. This model results in an high-energy tail on the electron distribution function and is of considerable interest in solar flares. The estimates presented by Miller *et al.* (1996) make the model a viable candidate in acceleration theory during flares.

Ion Acceleration. As Simnett pointed out, ions are the main carriers of energy in solar flares. Their acceleration is of major interest. Miller & Roberts (1995) have developed a model which nicely accounts for their stochastic acceleration in extended MHD-wave turbulence. The acceleration times are reasonable, and the evolution of a high energy tail (which is not a power-law) proceeds in a reasonably short time accompanied by a decrease in wave power. In addition to such stochastic acceleration models, ions can be accelerated by electric fields but into opposite direction as electrons. Of importance is ion acceleration at the reconnection site. Here, ions encounter an electric field during their meandering motion and may be accelerated to high energies when picking up the field energy. The acceleration time is equal to their transit time along the neutral line, and the energy gained is the electric potential drop along the neutral line. Since this is transverse to the magnetic field, it can be very large, leading to efficient acceleration (e.g. Lyons & Pridmore-Brown, 1996). Martens (1988) has provided a model of ion-beam generation in two-ribbon flares working in current sheets which is closely related to this mechanism.

Finally, a mechanism which provides transverse acceleration of ions is found in chaotic resonance of ions with transverse electric wave fields. This mechanism generates high temperatures and large anisotropies of ions.

4.2   Interplanetary Acceleration

Electron Acceleration. In interplanetary space, electron acceleration is restricted to stochastic acceleration in various wave fields. It is therefore believed to be of no importance in electrons of solar origin. Beams coming from the Sun suffer from scattering at self-excited plasma waves. If these waves undergo quasilinear diffusion in velocity space, the beam would be destroyed and the electrons heated. This process is not observed. At the contrary, beams seem to be relatively stable (e.g. Muschietti 1985). If the self-excited waves collapse, the distribution will evolve into a power-law tail. Current resolution does not provide sufficient information about the presence of the correct form of the power-law produced in these collapses.

Close to shock waves in interplanetary space, observations of reflected and diffuse electrons show that acceleration at shocks is at work (Wu 1984, Sonnerup 1968). Langmuir waves are generated along the electron path along the magnetic field, but no depletion of electrons is observed. Moreover, electron scattering as in the mechanism of Miller *et al.* (1996) has not been found. The electron acceleration seems to take place at the shock itself in the heating and reflection process in the shock front, with the electrons outside the shock degenerating into a halo distribution (Feldman *et al.* 1985). In the shock, ion- and electron-acoustic waves seem to be the most important electron scatterer.

Ion Acceleration. The most important ion accelerator in interplanetary space is second-order Fermi acceleration. Shock reflection including trapping of the ions in the shock front produces seed ions for further acceleration in interaction with their self-excited highly nonlinear MHD-waves. But heating and injection of ions also takes place in the shock front itself in interaction of the solar wind stream ions with large amplitude waves inside the shock transition. The main acceleration is experienced by the ions in the stochastic process in the foreshock wave field which is best described by a mechanism like that of Miller & Roberts (1995).

5   Open Problems

We list the most important open problems that became evident during our discussions in the Working Group and should be addressed in future studies:

Coronal Beams:

1. *Identification of acceleration mechanisms*, (stochastic, shocks, electric DC-field acceleration, etc.)
2. *Localization of acceleration sites* (inside versus outside [e.g. above] flare loops.)
3. *Role of magnetic reconnection* (Is it the main driver of flares? How does it control particle acceleration?)
4. *Fragmentation of acceleration region* (Spatial resolution of imaging instruments is still unsufficient.)
5. *Tracing of electron beams* (Are upward/downward directed electron beams accelerated together or independently?)

6. *Asymmetry of upward/downward acceleration*, (magnetic topology, escape probability, directionality of acceleration).
7. *Detection of ion beams* (no known signature in radio or HXR).
8. *Ion versus electron ratio* (of different acceleration mechanisms as function of energy).

Interplanetary Beams:

1. *Identification of acceleration mechanisms* (acceleration in flares, second-step acceleration, interplanetary shocks).
2. *Origin of interplanetary electron and ion beams*, (shocks, coronal mass ejections, magnetic reconnection sites, coronal versus chromospheric origin and elemental abundances [FIP-effect]).
3. *Evolution of velocity distribution of propagating beams* (high time resolution is needed to resolve positive slopes *in situ*).
4. *Modulation of interplanetary electron fluxes*, (variability of continuous or discrete coronal injections, MHD instabilities in flux tubes?)
5. *Role of ion beams*, (electron-to-ion ratios, relative energetics, detection methods for ion beams).

Acknowledgements This report grew out of the discussions and contributions of the participants in the working group: K. Arzner, D. Biesecker, A. Buttighoffer, M. Dellouis, G. Dulk, J. Hildebrandt, S. Hoang, M. Karlicky, M.R. Kundu, Y. Leblanc, R. Lin, D. Maia, M. Pick, M. Poquérusse, A. Raoult, W. Stehling, and V.Zaitsev. The authors of this report thank these colleagues for their lively and stimulating contributions. We thank A. O. Benz and R.P Lin for reading and commenting on the manuscript.

References

Antonucci E., Gabriel A.H., Dennis B.R. (1984): Astrophys. J. **287**, 917
Aschwanden M.J., Benz A.O., Schwartz R.A. (1993): ApJ **417**, 790
Aschwanden M.A., Benz A.O. (1995): ApJ **438**, 997
Aschwanden M.J., Benz A.O., Dennis B.R., Schwartz R.A. (1995): ApJ **455**, 347
Aschwanden M.J., Benz,A.O., and Tsuneta,S. (1997): ApJ, submitted
Aschwanden M.J., Hudson H.S., Kosugi,T. Schwartz R.A. (1996a): ApJ **464**, 985
Aschwanden M.J., Kosugi T., Hudson H.S., Wills M.J., Schwartz R.A. (1996c): ApJ **470**, 1198
Aschwanden M.J., Wills M.J., Hudson H.S., Kosugi T., Schwartz R.A. (1996b): ApJ **468**, 398
Bastian T.S., Vlahos L. (1996): (this volume)
Batchelor D.A., Crannell C.J., Wiehl H.J., Magun A. (1985): Ap. J. **295**, 258
Bentley R. (1996): (this volume)
Benz A.O. (1985): Solar Phys. **96**, 357
Benz A.O., Simnett G.M. (1986): Nature **320**, 508

Benz A.O., Lin R.P. (1996): (this workshop)

Benz A.O., Magun A., Stehling W., Su H. (1992): Solar Phys. **141**, 335

Biesecker D. (1996): (this workshop)

Brown J.C., Melrose D.B., Spicer D.S. (1979): Astrophys. J. **228**, 592

Brueckner G.E., Bartoe J.-D. F. (1983): Astrophys. J. **272**, 329

Buttighoffer A., Pick M., Hoang S. (1996): (this workshop)

Buttighoffer A., Pick M., Roelof E.C., Hoang S., Mangeney A., Lanzerotti L.J., Forsyth R.J., Phillips J.L. (1995): J. Geophys. Res. **100**, 3369

Chupp E.L. (1996): In *High Energy Solar Physics*, eds. R.Ramaty, N.Mandzhavidze, and X.-M. Hua, AIP Conf. Proc. **374**, p.3

Dere K.P., Bartoe J.-D. F., Brueckner G.E., Ewing J., Lund P. (1991): J. Geophys. Res. **96**, 9399

Doschek G.A., Feldman U., Kreplin R.W., Cohen L. (1980): Astrophys. J. **239**, 725

Dubouloz N., Treumann R.A., Pottelette R., Lynch K.A. (1995): Geophys. Res. Lett. **22**, 2969

Dulk, G.A., Steinberg J.L., Hoang S., Goldman M.V. (1987): Astron. Astrophys. **173**, 366

Feldman W.C. (1985): In *Collisionless shocks in the Heliosphere: Reviews of Current Research*, edited by. B.T. Tsurutani and R.G. Stone., AGU Geophys. Monogr. **35**, 195

Forman M.A., Ramaty R., Zweibel E.G. (1986): In *The Physics of the Sun, II*, edited by P.A.Sturrock, T.E.Holzer, D.Mihalas, and R.K.Ulrich, Reidel:Dordrecht, 249

Forrest D.J., Chupp E.L. (1983): Nature **305**, 291

Gosling J.T., McComas D.J. (1987): Geophys. Res. Lett. **14**, 355

Hildebrandt J., Stepanov A.V. (1997): Astron. Astrophys. (in preparation)

Hoang S., Dulk G.A., Leblanc Y. (1996): (this workshop)

Holman G.D. (1985): Astrophys. J. **293**, 584

Karlicky, M., Mann G., Aurass H. (1996): Astron. Astrophys., (in press)

Kopp R.A., Pneuman,G.W. (1976): Solar Phys. **50**, 85

Kundu M.R., Raulin J.P., Pick,M., Strong K.T. (1995a): Astrophys. J. **444**, 922

Kundu M.R., Raulin J.P., Nitta N., Hudson H.S., Shimojo M., Shibata, K., Raoult A. (1995b): Astrophys. J. **447**, L135

Kundu M.R., White S.M., Gopalswamy N., Lim J. (1994): Astrophys. J. Suppl. Ser. **90**, 599

LaBelle J., Treumann R.A. (1996): Geophys. Res. Lett. (in preparation)

Leblanc Y., Dulk G.A., Hoang S. (1995): Geophys. Res. Lett. **22**, 3429

Lin R.P. (1974): Space Science Rev. **16**, 189

Lin R.P. (1996): (this volume)

Lyons L.R., Pridmore-Brown D.C. (1995): In *Space Plasmas: Coupling Between Small and Medium Scale Processes*, (Geophysical Monograph 96, American Geophysical Union), p. 163

Maia D., Pick M., Hoang S., Manoharan P.-K. (1996): (this workshop)

Mandzhavidze N., Ramaty R. (1992): Astrophys. J. **389**, 739

Manoharan P.K., van Driel-Gesztelyi L., Pick M., Demoulin P. (1996): Astrophys. J. **468**, L73

Martens P.C.H. (1988): Astrophys. J. **330**, L131

Masuda S. (1994): PhD Thesis, University of Tokyo

Miller J.A., LaRosa T.N., Moore R.L. (1996): Astrophys. J. **461** 445

Miller J.A., Roberts D.A. (1995): Astrophys. J. **452** 912

Muschietti L. (1990): Solar Phys. **130**, 201

Nakajima H., Kosugi T., Kai K., Enome, S. (1982): Nature **305**, 292

Pick M., Buttighoffer A., Kerdraon A., Armstrong T.P., Roelof E.C., Lanzerotti L.J., Simnett G.M., Lemen J. (1995 b): Space Sci. Rev. **72**, 315

Pick M., Lanzerotti L.J., Buttighoffer A., Hoang S., Forsyth R.J. (1995 a): Geophys. Res. Lett. **22**, 3377

Pick M., Lanzerotti L.J., Buttighoffer A., Sarris E.T., Armstrong T.P., Simnett
G.M., Roelof E.C., Kerdraon A. (1995 b): Geophys. Res. Lett. **22**, 3373

Poquérusse M., Hoang S., Bougeret J.-L., Moncuquet M. (1995): In *Solar Wind
8,*(Proc. Conf. in Dana Point, California, June 1995, in press)

Ramaty R., Mandzhavidze N., Kozlovsky B., and Murphy R.J. (1995): Astrophys.
J. **455** L193

Reiner M.J., and 10 co-authors (1995): Space Sci. Rev. **72**, 261

Retterer J.M., Chang T., Jasperse J.R. (1995): In *Space Plasmas: Coupling Between Small and Medium Scale Processes,* (Geophysical Monograph 96, American
Geophysical Union), p. 127

Roelof E.C., Gold R.E., Simnett G.M., Tappin S.J., Armstrong T.P., and Lanzerotti L.J. (1992): Geophys.Res.Lett. **19**, No. 12, 1243

Sakao T. (1994): PhD Thesis, University of Tokyo

Shibata K., Masuda S., Shimojo M., Hara H., Yokoyama T., Tsuneta S., Kosugi
T., and Ogawara Y. (1995): Astrophys. J. **451**, L83

Simnett G.M., Haines M.G. (1990): Space Sci. Rev. **130**, 253

Simnett G.M. (1994): Space Sci. Rev. **70**, 69

Simnett G.M. (1995): Space Sci. Rev. **73**, 387

Simnett G.M. (1986): Solar Phys. **106**, 165

Sonnerup B.U.Ö. (1986): J. Geophys. Res. **74**, 1301

Thomsen M.F., Gosling J.T., Bame S.J., Russell C.T. (1990): J. Geophys. Res. **95**,
957

Thomsen M.F. (1985): In *Collisionless shocks in the Heliosphere: Reviews of Current Research,* edited by. B.T. Tsurutani and R.G. Stone. (Americ. Geophys.
Union, Geophys. Monogr. 35), 253

Treumann R.A. and 12 co-authors (1986): Adv. Space Phys. **6**, 93

Treumann R.A., LaBelle J. (1996): (this workshop)

Tsuneta S. (1995): Publ. Astron. Soc. Japan **47**, 691

Tsuneta S., Hara H., Shimizu T., Acton L.W., Strong K.T., Hudson H.S.,
Ogawara Y. (1992): Publ. Astron. Soc. Japan **44**, L63

Vlahos L. (1993): Adv. Space Res. **13** No.9, 161

Wild J.P., Roberts J.A., Murray J.D. (1954), Nature **173**, 532

Wu C.S., Gaffey J.D., Jr., Liberman B. (1981): J. Plasma Phys. **25**, 391

Wu C.S. (1984): J. Geophys. Res. **89**, 8857

Yokoyama T., Shibata K. (1996): Publ. Astron. Soc. Japan **48**, 353

Part III

Large-Scale Disturbances

Coronal Mass Ejections and Type II Radio Bursts

Henry Aurass

Astrophysikalisches Institut Potsdam, Observatorium für solare Radioastronomie, Telegrafenberg A31, D - 14473 Potsdam, Germany

Abstract. Coronal mass ejections (CMEs) and shock wave induced radio bursts (type II) are reviewed. CMEs are - beneath flare blast waves - invoked to be the drivers of type II burst emitting super-Alfvénic disturbances. The paper focuses on the available experimental evidence for this assumption. For instance, the apparent contradiction is discussed between measured speeds of potential shock drivers and their observed type II burst association. Further, several examples are presented for recognizing CMEs by the appearance of other characteristic nonthermal radio signatures. Open problems are assembled which should be newly attact by high time and frequency resolution decimeter and meter wave radio spectral and imaging observations from ground combined with visible light and X-ray imaging data from YOHKOH and SOHO space experiments. The paper shows that radio observations in general, and especially the radiation of shock accelerated electrons, constitute a unique access to the structure and the dynamics of the coronal magnetoplasma. This is important for understanding the timing and the sites of energy release processes in the solar corona and for studying the physics of collisionless shock waves in space plasmas.

Résumé. Nous présentons une revue des caractéristiques observationnelles des éjections de matière coronale (EMC's) et des ondes de choc dont les signatures sont des sursauts radio de type II. Il est généralement admis que EMC's et ondes de choc explosives, associées aux éruptions, sont à l'origine des perturbations super-alfveniques qui engendrent les type II. Touutefois il existe un désaccord apparent entre les vitesses mesurées pour EMC's et ondes explosives et celles des sursauts de type II. Magré cela, nous montrons sur des exemples que certaines émissions radio non-thermiques constituent des signatures des EMC's. De plus nous identifions un ensemble de problèmes pour lesquels la combinaison d'observations radio decimétrique-métrique, obtenues au sol avec de hautes résolutions temporelle, spectrale et spatiale, avec l'imagerie en lumière blanche et en rayons X, fournie par des instruments embarqués sur les satellites SOHO et YOHKOH, devraient permettre de mieux cerner. Plus généralement nous tentons de montrer que les observations radio, et plus particulièrement celles générées par les électrons accélérés par les chocs, fournissent un moyen unique pour étudier la structure et la dynamique du magnéto-plasma coronal. Ceci est important pour: (i) caractériser l'évolution temporelle et localiser les sites de la libération d'énergie dans la Couronne; (ii) étudier la physique des ondes de choc non collisionnelles dans les plasmas naturels.

1   Introduction

Solar flares and solar coronal mass ejections - CMEs - (or coronal magnetic ejections, Hundhausen 1995) are energetically the most important transient phenomena of the solar atmosphere. Both can initiate - either as an expanding blast wave or as an escaping massive piston - a moving MHD-like disturbance which grows in the solar corona (and/or later in the solar wind) to a collision-less shock wave if its speed is larger than the local Alfvén velocity in the background atmosphere. In the meter wave range (300 - 30 MHz) radio spectral observations reveal slowly drifting features (type II bursts). This is usually taken as evidence of super-Alfvénic disturbances. Type II bursts are then the radio signature of energetic electrons (up to tens of keV) accelerated at flare or CME related shock waves. It should be emphasized that radio emission of shock-accelerated particles is the only ground based observational access to coronal shock waves. From spacecraft, in-situ observations of shock front passages and remote sensing of low frequency (hectometer) shock-associated radio emission are possible in the solar wind (see e.g. Bougeret 1985 for a review).

There is a gap between ground and space based radio observations of type II bursts caused by the ionosphere, the probably low intensity of shock related radio emission below 20 MHz, and the high radio background in this range. Only the recently flown CORONAS I and WIND/WAVES experiments provided a sufficient overlap with the frequency range accessible from ground. However, an example of a continuous tracing of a radio visible shock from meter waves down to hectometer waves has not yet been reported in the literature. It is well possible that coronal and interplanetary radio shock signatures must not appear together, and could have different drivers within one solar event (especially in the case of a flare associated CME).

Shocks and CMEs are quoted by the solar energetic particle community as accelerators of MeV electrons and protons, but the details of how to get such high energies under coronal conditions are not yet properly understood (Mann, pers. comm.). Notice that Klein et al. (1988) and Klein et al. (1995) have shown that extended coronal shocks play a minor part in the acceleration of relativistic electrons observed in the low corona.

At present two solar oriented space missions offer a new quality of imaging observations of the solar corona:

- The soft X-ray telescope (SXT) onboard the YOHKOH mission (first data September 1991) presents high cadence images (Tsuneta et al. 1991) of hot and dense coronal magnetoplasma structures;
- The SOHO mission (first data April 1996) includes the first orbiting coronograph (LASCO, Brückner et al. 1995) working down to a height of 0.1 $R_\odot$ above the photosphere.

Both these new space experiments provide detailed information about struc-ture and motions of the coronal magnetoplasma. Both cover the same height range in which the nonthermal decimeter and meter wave radio burst sources are located. This provides a new opportunity to investigate the potential of existing ground based solar observations which, in the contrary, are sensitive to nonthermal electrons energized during transient solar phenomena.

There is a large amount of previous work about solar radio type II bursts on the one hand and about the relationship of radio phenomena with the appearance of CMEs in white light observations on the other hand (for a summary see e.g. Hildner et al. 1986, and Section 2 of the present paper). Definitive results about radio phenomena associated with CMEs are mainly obtained by radio imaging facilities at low frequencies - the Culgoora Heliograph (160, 80, 43 MHz, for an instrumental description see Labrum 1985), the Clark Lake Radioheliograph (120 - 25 MHz, see Kundu et al. 1983), and the Nançay Radio Heliograph - NRH - (The Radioheliograph Group 1993). At present, the NRH is the only working meter wave imaging facility dedicated to the sun. A complete review of all these radio physical results is beyond the scope of this paper. Published work shows that the observation of shock- and CME related radio phenomena requires simultaneous high time and spectral resolution in the whole meter and decimeter range combined with imaging data of comparable time resolution on at least two frequencies within the range of spectral observations. In Europe, the NRH working at five frequencies with 0.1 s time resolution combined e.g. with the Potsdam-Tremsdorf 40 - 800 MHz spectrometer system (Mann et al. 1992) are best fitting the given instrumental demands. During the flight of YOHKOH and SOHO this gives a promising perspective for association studies.

In Section 2 some aspects of CMEs and of the radio signature of coronal shock waves are briefly reviewed. In Section 3 some answers are assembled to the question "How to see a CME in radio?".

2   Summary About CMEs and Coronal Shock Waves

In recent years several reviews concerning CMEs have been published. We refer to Harrison (1991), Kahler (1992), Steinolfson (1992), Chertok (1993), Low (1993), Webb (1994), Dryer (1994), Hundhausen (1994), Webb et al. (1994), and Hundhausen (1995). For reviews about the radio signature of coronal shock waves see e.g. Bougeret (1985), Aurass (1992), and Mann (1995a). This paper focuses on common aspects of CMEs and coronal shock waves and on nonthermal radio signatures of both phenomena.

2.1   Coronal Mass Ejections

Morphology. A coronal mass ejection appears on white light coronograph images as brightening, blowing up and ejection of a more or less extended region of the solar corona. Despite the fact that CMEs have been observed for 20 years they are not yet fully understood. CMEs reveal a large scale

138   Henry Aurass

evolutionary process of reconfiguration of the coronal magnetic field (Sime 1989).

Burkepile and St. Cyr (1993) have shown impressive CME images and have reported statistical information about the different CME forms. One third of all CMEs consists of the typical structural elements of the loop or bubble type. The outermost feature is a bright leading arch, followed by a dark cavity and a bright core of dense matter. The bright core is usually identified with parts of an uplifting filament, the dark cavity might be the disconnected coilshaped flux system which was supporting the filament before. There are also such CMEs without a bright core. About a quarter of all CMEs consists of a diffuse brightness enhancement (type "material" or "cloud"). A further 25 % of CMEs have very different but well defined forms. According to Kahler (1991) "...typical CMEs tend to occur in streamer structures and destroy or significantly modify these structures ...". Only 1.9 % of all CMEs are themselves formed like streamers. Coronal helmet streamer latitudes and CME latitudes have coincident scatterplots over the solar activity cycle (Hundhausen 1993). But the presence of a streamer structure is not necessary for CME formation. For near-minimum conditions Kahler (1991) found CMEs associated with an active region without an overlying streamer structure, and streamers over active regions without a CME.

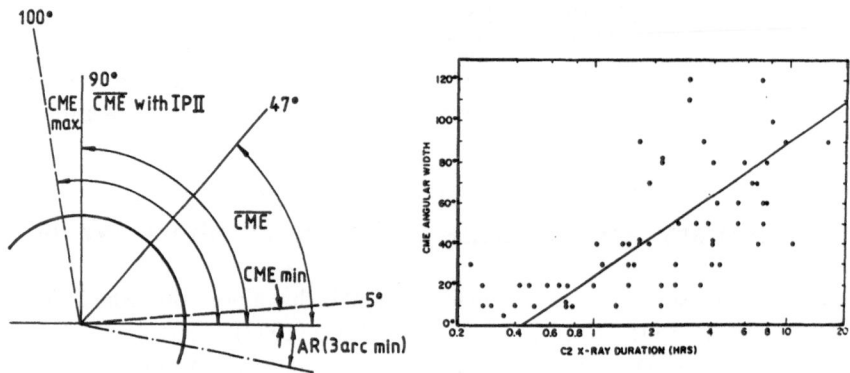

Fig. 1. a - left: The spatial scale of CMEs acording to Hundhausen (1993) and Cane et al. (1987). AR means active region. b - right: The linear relation between the log. duration of an soft X-ray burst and the span of the associated CME (from Kahler et al. 1989).

Figure 1a summarizes the information about the angular span of CMEs as given by Hundhausen (1993) and using Cane et al. (1987). Three facts are of special importance for analyzing associations between CMEs and flare (active region) related phenomena:

- The angular extent of an average CME is nearly five times larger than a medium size active region.
- The smallest CMEs have about the same span as an active region.
- Those CMEs which are associated with a low frequency interplanetary type II burst are on average twice as large as a typical CME.

Association with Soft X - Ray Phenomena. There is a statistically confirmed relationship between the angular extent of a CME and the duration of the simultaneously observed soft X-ray burst (Kahler et al. 1989 and our Figure 1b). This means on average the spatial scale of a CME determines the time scale of an associated energy dissipation process (e.g. heating) in the corona. Hundhausen (1995) points the attention on some exceptional examples.

What is known about the time sequence and the spatial relation of CMEs and associated X-ray sources ?

Harrison et al. (1990) report on minor X-ray events leading in time the CME onset and the onset of the associated main X-ray burst. The time interval between minor and main X-ray event is of the order of tens of minutes. This is the same time scale as given for early nonthermal energy release signatures in microwaves (Kai et al. 1983) and decimeter waves (Averianikhina et al. 1990).

One of the discoveries of the YOHKOH mission results from SXT's ability to see CME related features on the disk. Hiei et al. (1993) report on coronal mass loss observations during a high latitude arcade event. Hudson (1995) describes a structured soft X-ray cloud adjacent to a flare on the disk which starts moving and disappeares completely within 1 hour. McAllister et al. (1994, 1996) present evidence for a polar crown CME which passed by the Ulysses spacecraft and caused a strong geomagnetic storm. No flare and evident soft X-ray flux change, nor a big prominence disappearance was noticed. Within 10 hours after the probable solar ignition of the event, a faint but very elongated (150 degs in longitude and 30 degs in latitude !) X-ray arcade was formed between coronal hole boundaries and above a highly warped magnetic neutral line. The only feature resembling to flare activity were He 10830 ribbons at the arcade footpoints.

In YOHKOH images, Švestka et al. (1995) have recognized hints on to the process of arcade formation in its vertical extent - the long duration growth of giant X-ray arches. These arches were already discovered in SMM X-ray images. Švestka et al. (1995) argue from the analysis of 7 events that the expanding giant arches are a consequence of (or at least associated with) CMEs. In extension of the H_α post flare loop evolution, the giant arches grow with constant speeds between 1.1 and 12.1 $km s^{-1}$. They represent relatively strong density enhancements which slowly decay with time. For one example, there is a fivefold density decrease reported over a time interval of 11.5 hours which corresponds with a giant arch volume expansion rate of 0.7 % per minute. We come back to this result in Section 3.

The observations point to the relevance of Kopp and Pneuman's (1976) model of field line reclosure due to magnetic reconnection after prominence eruptions for understanding CMEs. For a detailed comparison of soft X-ray and white light coronal features it is important that the LASCO system onboard SOHO will cover the same height range in which the soft X-ray giant arches have been observed. This is in contrast with the SMM coronograph.

Other Associations. The discussion about the spatial scales of CMEs is a suitable point to ask generally for associations of CMEs with other classes of solar activity. Webb and Howard (1994) reanalysed the data of all orbiting coronographs. They found that no one class of solar activity is better correlated with the CME rate over the solar activity cycle than any other.

An important point is the relationship of CMEs to flares and to active regions. As already noted (Figure 1a) the angular span of an average CME is at least 5 times larger than active region magnetic field structures in the corona. Only the smallest CMEs start on spatial scales which are comparable to flare structures. The association of a CME with an erupting active region filament is simple only in cases of spatially narrow CMEs. If a broad CME is associated with a flaring active region it can be situated anywhere under the arch spanned by the CME (Harrison et al. 1990). Kahler (1991) found under minimum solar activity conditions equator-crossing CMEs rooted in two active regions.

As already mentioned a CME can be accompanied by the disappearance of quiescent filaments (e.g. Mouradian et al. 1996) and by flares. Not all features existing in the disturbed corona behind the leading parts of the white light CME need to be visible in white light images, some of them - e.g. the erupting filament, the moving type IV burst, the magnetic field structures in the lower corona - are discussed in the flare context, too. This may be a source of terminology confusion.

2.2   Coronal Shock Waves

All experimental facts about shock waves in the solar corona result from remote sensing in the meter wave range. This is possible due to the fact that coronal shocks accelerate electrons to energies of several tens of keV leading to an enhanced level of high frequency plasma wave turbulence. The high frequency (Langmuir or upper hybrid) waves can be scattered off ions or off low frequency plasma waves (e.g. ion sound). Further, they can coalesce with other high frequency waves. Thus one gets radio emission of type II near the local plasma frequency (f_p, F) and/or its second harmonic ($2 f_p$, H). For type II burst terminology see Nelson and Melrose (1985) and Aurass (1992).

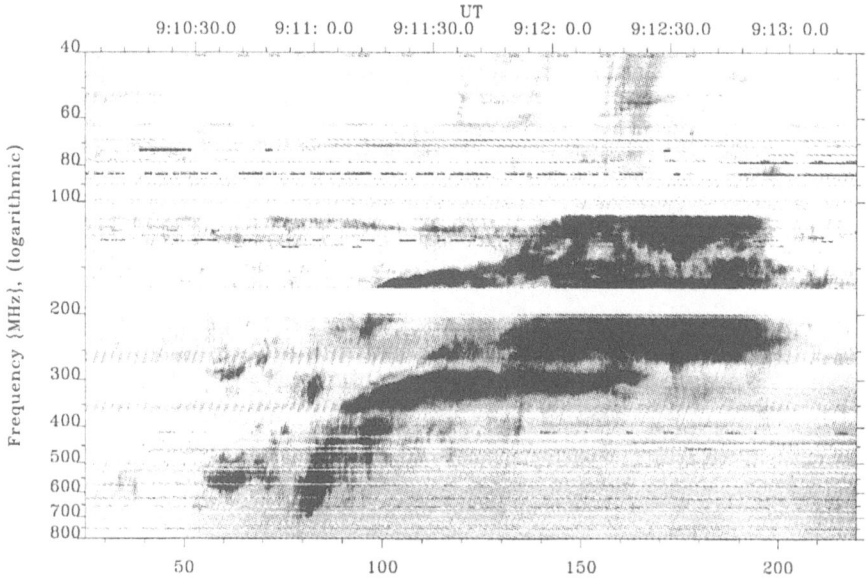

Fig. 2. The type II burst of 09 July 1996.

Spectral Data. Figure 2 shows a digitally processed type II burst spectrogram (background subtracted). Stripes parallel to the time axis are terrestrial transmitters. Note the high starting frequency (350 MHz for the F lane) and the complex and multiple lane pattern. Figure 3 presents the fundamental lane of a type II burst with extremely strong herringbone structure. There is also a backbone interval without herringbones (12:11:22 - 35). This reveals that the backbone is an independent component and not a superposition of herringbones.

Aurass et al. (1994a) find several examples for type II bursts with three harmonically related lanes. Together with the usual F and H lanes there appears a lane at thrice the plasma frequency ($3 f_p$). After early reports (Aurass 1992 and references therein) it is now a well-proven experimental fact that type II emission at the third harmonic of the plasma frequency is not as rare and doubtful as believed earlier. An extended investigation (Zlotnik et al., this conference) reveals a brightness temperature ratio from 4 to 1000 between the H and the third harmonic source.

According to common knowledge (e.g. Švestka 1976) the ignition of the type II emitting disturbance can drive a Moreton wave in the photosphere. The launch time of flare associated shocks was newly analysed in a statistical study about type II bursts starting at high frequencies (Vršnak et al. 1995). The result favours a launch time prior to the first peak of the associated microwave burst. Karlický and Odstrčil (1994) describe type II burst related

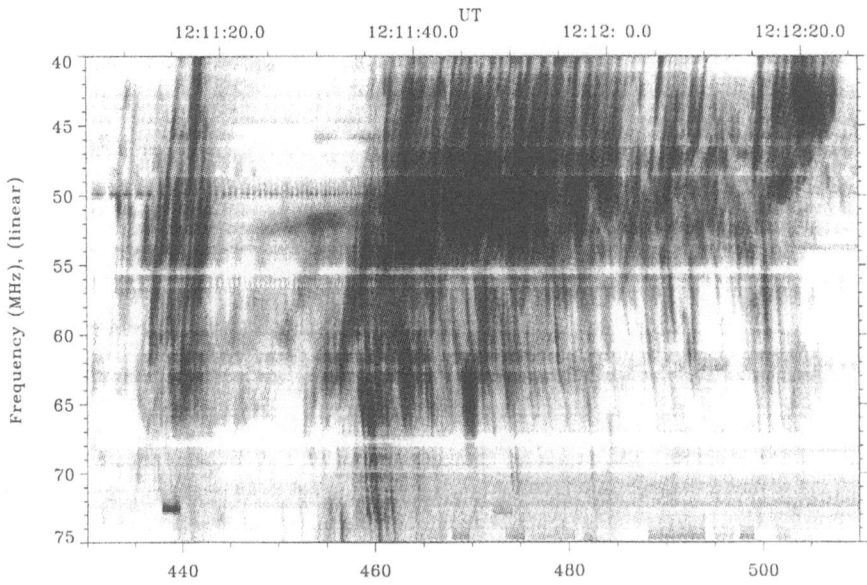

Fig. 3. The type II burst of 30 June 1995 with strong herringbone fine structure in the fundamental lane.

features seen above 1 GHz. Aurass et al. (1994b) study a typical pattern of decimetric emission with a triangular envelope (a trianglar spectral pattern, TSP) in the radio spectra which was found to systematically preceede some type II bursts and becomes possibly several times visible in the spectrum with delays of 3 to 4 minutes between. In Figure 4 we show two examples of this effect. Here, the TSP together with some type III bursts is the earliest feature of the event to precede the impulsive microwave and γ-ray burst. In this example the second TSP immediately precedes the type II burst lanes. The data support the idea that for some flares the type II burst emitting disturbance is initiated by energy release processes in large coronal heights. Klassen (1996) presents a TSP which interferes with a superposed pair of a type III and a reverse drift burst appearing in absorption on the TSP background. This leads to the conclusion that this TSP was exited near a cusp-shaped magnetic field structure. Tsuneta (1996) has recently emphasized the importance of cusp-shaped structures for the flare process.

In studying type II bursts several questions arise, e.g.

- What is the thickness of the shock transition region under coronal conditions ?
- What is the site of radio emission in the shock reference frame ?
- Where are the sites of particle acceleration at the shock front ?

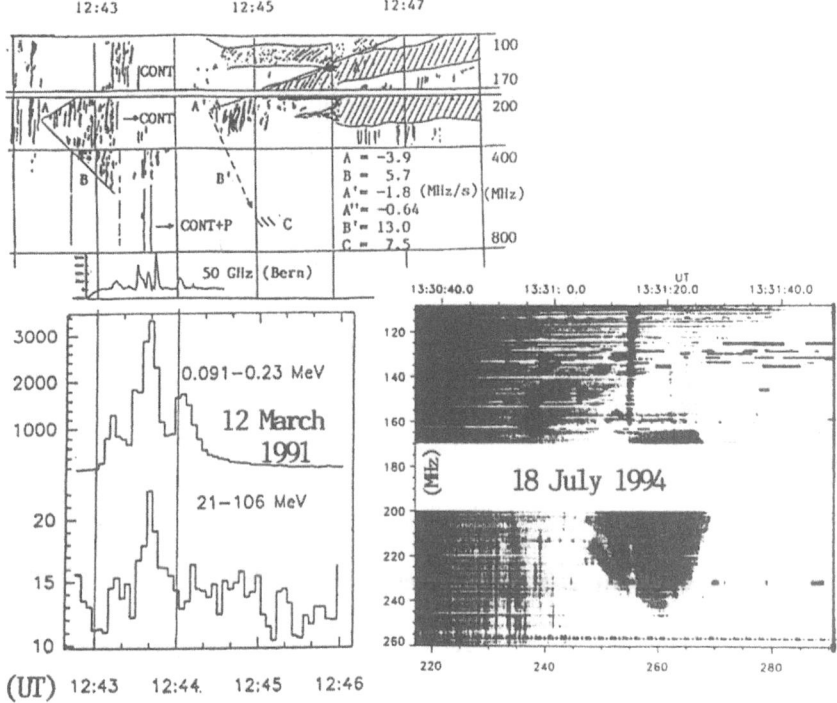

Fig. 4. Two examples of triangular spectral patterns. Left: Event 12 March 1991, top: sketch of the dynamic spectrum observed at Potsdam-Tremsdorf. Middle: the microwave burst (from Aurass et al. 1994b) and (bottom) the high energy emission (Vilmer 1993) begin 30 s past the first TSP. The second TSP precedes the type II lanes. Right: the TSP preceding the 18 July 1994 type II burst (from Klassen 1996).

The transition region of a collisionless shock has a thickness of several ion inertial lengths (cf. Mann et al. 1995 and references therein). The ion inertial length is a plasma parameter (Krall and Trivelpice 1973) ranging from about 7 m at a plasma frequency level of 300 MHz (low corona) to about 100 km (plasma frequency level of 20 kHz, solar wind plasma at 1 AU). This means the shock transition region of coronal disturbances is a thin surface of some tens of meters width which cannot be resolved by radio imaging instruments. We assume that fast mode shocks are the most probable source of type II bursts among those possible in the corona (Mann et al. 1995). Then its transition region is characterized by a positive jump of the magnetic field strength, the temperature and the density. The mean instantaneous

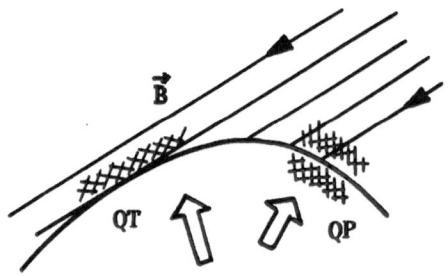

Fig. 5. Scheme of a shock wave with quasi-parallel (QP) and quasi-transversal (QT) transition intervals. Regions of particle acceleration are cross-hatched (according to Mann and Claßen 1995).

bandwidth of type II bursts (about 30 %, Mann et al. 1996) leads to an average density jump of 74 % of the background. The shock transition region represents a strong density jump which becomes visible as plasma emission of shock accelerated electrons. The frequency of plasma emission depends on the density - therefore a highly resolved spectrum in time and frequency offers a look into otherwise unobservable spatial scales. Of course it is impossible to read this information without theoretical guidelines.

From in-situ observations of shock waves in the interplanetary space (Mann et al. 1995 for references) it is known that high frequency plasma waves are mainly concentrated in the upstream region. Low frequency plasma waves dominate in the downstream region. Since both components are necessary for the formation of escaping radio emission Mann et al. (1995) conclude that the radio source region must be situated at the overlap of both components - at the shock transition.

In Figure 5 we have sketched a shock front and the upstream field lines of the background magnetic field. "QT " is a quasi-transversal shock, "QP " characterizes a quasi-parallel shock transition. The shock drift acceleration process (see Holman and Pesses 1983, Benz and Thejappa 1988) acts immediately in front of the QT transition region (cross-hatched in Figure 5). The energy of the accelerated electrons depends sensitively on the angle between the shock normal and the magnetic field. For getting the necessary energies to explain the backbone emission, but much more for getting particle speeds consistent with herringbone data (e.g. Mann and Klose 1995) a very sharp angular criterion has to be fullfilled. For the observer, it looks strange to assume that during several minutes of type II burst duration the mentioned angle should persist e.g. between 88 and 90 degrees, given the structural complexity of the coronal magnetoplasma. This problem has been resolved recently by

Mann et al. (1994) and Mann and Claßen (1995) presuming that the same basic physical processes act at collisionless shocks in different space plasmas. They assume that QP shock waves are - in the same manner as at the earth's bow shock - represented by an ensemble of short large amplitude magnetic field structures (SLAMS, cross-hatched in Figure 5) appearing in the upstream and the downstream region. SLAMS have an amplitude-dependent relative speed in the shock reference frame (Mann, 1995b). Therefore they can act as moving magnetic mirrors. Mann and Claßen (1995) show that also at QP shock waves electron acceleration is possible up to high energies.

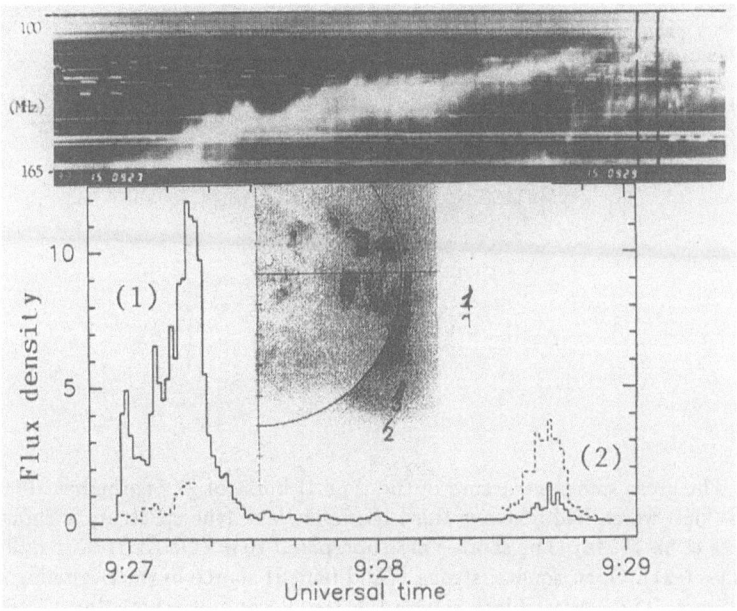

ig. 6. An example for the type II burst large scale source structure. For explanations see text.

Imaging Data - the Source Structure of Type II Bursts. Early reports from Culgoora observations deal with large ($\gtrsim 0.5~\mathrm{R}_\odot$) sources consisting of different archlike distributed subsources (Nelson and Melrose 1985 for detailed reference). According to recent experience it is impossible to judge about the source structure without an exact spectral identification. For illustration we present in Figure 6 a simple example (from Aurass et al. 1994a). The type II burst spectrum shows a single lane (probably the H lane, Figure 6 top between 100 and 165 MHz). The lane consists of two split bands with variable split bandwidth. The high frequency split band enters the given spectral range nearly 1.5 minutes later than the low frequency split band. This offers the opportunity to discriminate clearly the radio images of both split bands at the 164 MHz NRH observing frequency. The middle part of Figure 6 shows the corresponding source distribution. Only the source centers have been drawn for simplicity.

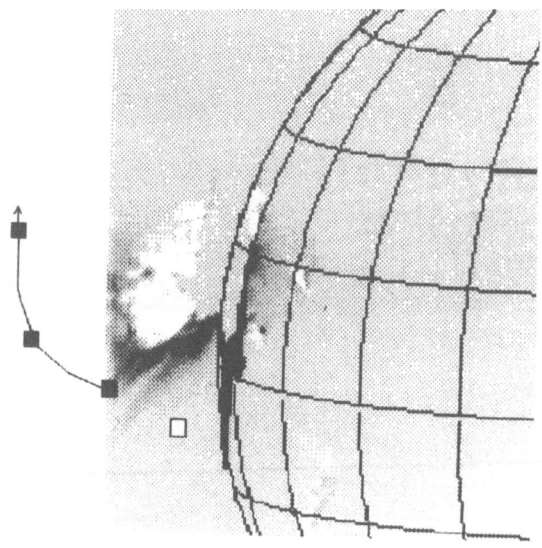

Fig. 7. The gross source structure of the type II burst of 27 September 1993 near the west limb which had a strong third harmonic lane (the spectrum is shown e.g. in Aurass et al. 1994a). The sources are superposed to a YOHKOH SXT difference image (see text). Open square: strong F and faint H source in the beginning of the type II burst. The nearest black square (on the X-ray image boundary): strong H and faint F as well as 3F source in the beginning of the type II burst. Further black squares: trace of the main H and the 3F source later in the burst.

The radio sources appear pairwise for each split band with about 0.5 $R_\odot$ distance between the source sites. They are superposed on to a YOHKOH SXT image and are situated on both sides of a large soft X-ray loop structure. During the passage of both split bands over the 164 MHz plasma level the NRH indicates a simultaneous brightening at both source sites with different weights. In the bottom of Figure 6 the flux curves (in arbitrary units for both sources 1, 2) are given. Recall that the band splitting reflects local properties of the wave front. It allows us to infer the density jump in the shock transition region which is a distance of some tens of meters under coronal circumstances. On the other hand both split bands in the spectrum are formed from contributions of widely separated source sites. We argue that there is a common electron population propagating along the shock front and / or along the field lines of the loop seen in soft X-rays. The changing intensity weight of both sources for the different split bands can be a directivity effect for electron streams or radio emission. This finding demands for a systematic investigation.

The discussed simple example confirms that for understanding more complex type II spectra combined high spectral and time resolution data are inevitably necessary for associating the (seemingly) complex spatial pattern of subsources with the correct spectral structure. Further, we stress that in the case of a somewhat lower sensitivity of the imaging instrument one would see for each split band one source, only (No. 1 for the low frequency split band, No. 2 for the high frequency split band, comp. Figure 6) with a large distance between. It is well possible that this finding explains some of the earlier reported discrepancies about the relative position of different spectral features of type II bursts (Dulk 1982, Nelson and Melrose 1985).

Aurass et al. (1994a) have first shown imaging data of a type II burst at three harmonically related frequencies (Figure 7). For the F and H lanes again a double source has been found for comparable spectral substructures. For this event, a clear image discrimination between the split band lanes is not possible. If measured at the same time on different frequencies the stronger F source is at the same site as the fainter H source and vice versa. The third harmonic source is single and appears at the same site as the main H source. These facts are consistent with the fundamental/harmonic interpretation and argue against significant effects of wave propagation on the radio emission at the analysed frequencies. Given the spectral extent of the event in relation with the NRH spectral coverage, in Figure 7 the sources can be shown initially for the F and the H lane, in later time intervals for the H and third harmonic source, only. Both move together on an appearantly archlike azimuthal path.

The radio sources are superposed on to a YOHKOH SXT difference image taken with the AlMg filter, 5 s exposure time, from the time before and after the flare (12:37 - 12:00 UT, without correction for solar rotation). The radio sources are all situated in a region which is hotter and denser after the flare.

Relation between Type II Source Positions and White Light CMEs.
The large scale shock structure along the shock transition is of principal interest in searching for magnetoplasma structures in the solar corona. It is also important for recognizing the driving agent (blast wave or piston) of the shock and might be connected with the microphysical aspect of the processes in the shock front (acceleration processes, directivity of electron streams). Simultaneous observations of CMEs and the position of type II bursts are

the only way to determine the spatial relationship between both effects. Further this provides two independent approaches for density estimation in the disturbed coronal plasma.

Despite all efforts there were no simultaneous observations during the SKYLAB experiment (Hildner et al. 1986). Gopalswamy and Kundu (1992a) counted "... a dozen CME events with simultaneous radio and optical coverage". Therefore it is not surprising that there is no clear and unique picture about the CME / type II relation until now. Gary et al. (1985) present a type II burst with sources below the top of loop structures within a CME. Gergely et al. (1984) describe a CME with a type II burst source moving tangentially along CME loop structures toward the solar disk. Kundu and Gopalswamy (1992) report on a type II source in front of a CME, and Gopalswamy and Kundu (1989) discuss a clearly flare related type II burst behind a CME. A flare induced disturbance launching during a possibly associated CME in progress seems to be best fitting the coronal type II burst data. In contrast there is evidence for low frequency type II bursts in the interplanetary space to be driven by the most massive, energetic and fastest CMEs acting like pistons (Cane et al. 1987).

2.3   The Speed Problem

The apparent speeds of CMEs (Hundhausen et al. 1994) range from 10 to 2100 $\mathrm{km\,s}^{-1}$ with a mean value of 349 $\mathrm{km\,s}^{-1}$ (all features) and 445 $\mathrm{km\,s}^{-1}$ (only the "outer loop feature"). The average apparent speed of type II bursts is about 700 $\mathrm{km\,s}^{-1}$ (Robinson 1985). For the transformation of the drift rate (the measured quantity in the spectral records, $D_f = df/dt$) in the exciter speed v_{exc} one implies a density model $N_e(s)$ and an assumption about the angle between the velocity vector and the density gradient according to

$$D_f = \frac{df}{dt} = \frac{1}{2\pi} \cdot \frac{d\omega_{\mathrm{pe}}}{dt} = \frac{f}{2} \cdot \frac{v_{\mathrm{exc}}}{L} \cdot \cos\left(\frac{dN_e}{ds}, v_{\mathrm{exc}}\right) \qquad (1)$$

Here $\omega_{\mathrm{pe}} = \sqrt{4\pi e^2 N_e/m_e}$ is 2π times the plasma frequency, L is the coronal density height scale, m_e is the electron mass, N_e is the electron number density, and e is the elementary charge.

Although it seems to be trivial, we note that low drift rates need not mean low exciter speeds. Characterized by the typical F/H pattern in the spectrum it is easy to find type II burst lanes with zero drift rate. Urbarz (pers. comm.) found 27 out of more than 400 type II bursts with drift rates ≤ 0 in the Weissenau spectral data between 1974 and 1987. Already Weiss (1963) and Gergely et al. (1984) have drawn the attention on very low drift rate type II bursts with large tangential source motions. Zero drift spectra reveal a disturbance growing to a shock wave only on that part of its path through the coronal plasma which is perpendicular to the coronal density gradient. Simplifying one could use the term azimuthal propagation in contrast with the mostly implied radial propagation.

A travelling collisionless MHD disturbance grows to a shock if the Alfvén speed of the background medium is smaller than the velocity of the disturbance. The local Alfvén speed v_A depends on the ratio of the magnetic field strength to the square root of the density - the disturbance grows to a shock in regions of $v > v_A = 2.1 \cdot 10^6 \cdot B/\sqrt{N_e}$ (B in 10^5 nT, N_e in cm^{-3}). Characteristic values for the Alfvén speed are about $500\,\mathrm{km\,s^{-1}}$ in the lower corona (at the 300 MHz plasma level assuming $B = 10^6$ nT) and $60\,\mathrm{km\,s^{-1}}$ in the solar wind at 1 AU (at the 20 kHz plasma level assuming $B = 6$ nT). In the solar wind the disturbance must additionally overcome the background streaming velocity of $470\,\mathrm{km\,s^{-1}}$ (mean value). Assuming the CME drives the disturbance, only fast CMEs should be associated with a type II burst. But this is not true: Gopalswamy and Kundu (1992) present a list of 6 cases of slow CMEs (speeds below $400\,\mathrm{km\,s^{-1}}$) accompanied by a type II burst in the corona. On the other hand, Sheeley at al. (1984) show a histogram of 34 CMEs being - despite a broad velocity scatter between 200 and $1600\,\mathrm{km\,s^{-1}}$ - without metric type II bursts. These facts underline and confirm that the appearance of a type II burst is not only determined by the disturbance velocity. There must be strong differences in the Alfvén speed of the plasma surrounding the source of the disturbance. This is a way to understand that in some flares the metric type II burst appears suddenly at relatively low frequencies - this would then only reflect the inhomogenity around the exciter and not a large height in the corona.

As proposed already by Uchida (1974) channeling of the disturbance appears into low v_A regions if the background Alfvén speed is inhomogeneous. The results of a statistical study by Person et al. (1989) show that the appearance of a coronal type II burst does not depend on the energy of the exciting flare and therefore point in the same direction. Aurass and Rendtel (1989) compare the type II burst productivity of two very flare active regions. They find that a coronal hole - with low Alfvén speed inside - in the immediate surroundings of one region could be the cause for the significantly enhanced number of type II bursts induced by the flares in this region. There is still another observational argument. Recall Figure 1a - those CMEs which are associated with a kilometric type II burst in the solar wind are on average twice as large as the average CME. Presuming an inhomogeneous solar wind a large angular span of the disturbance enhances the probability to meet low v_A regions. This is the effect which Kahler et al. (1984) could not find in the lower corona.

Summarizing: type II bursts can be the radio illumination of low v_A regions of the inhomogeneous corona independently from driving the shock by a piston (e.g. a CME) or getting the shock from a semi-spherically escaping flare blast wave. The relatively sparse data about the position of the

CME's leading edge to the type II burst sources seem to be in favour of a flare excited driver of the coronal shock wave. The same can be argued from the average CME - flare - type II onset timing. This conclusion is already summarized in Wagner and MacQueen's (1983) scheme: a flare induced disturbance illuminates low Alfvén speed regions in the legs of a CME in progress thus yielding a type II burst. The disturbance can later on overtake the CME. In contrast with this picture for coronal type II bursts, Cane et al. (1987) present evidence that radio emissive shocks in the interplanetary space are probably driven by energetic CMEs acting as pistons.

3   How to See a CME by Radio Methods?

The radio detection of CMEs is a largely unexplored territory (Bastian, this volume). It is an attractive challenge to detect CMEs by radio methods not only near the limb but also on the disk. There are only a few events studied commonly in radio and coronographic images due to the complexity of the problem and the lack of simultaneously working appropriate coronographs and radio telescopes. Available data mostly suffer from insufficient time and spectral coverage. Here we intend to summarize direct and indirect radio signatures of CMEs as follows:

- A direct hint for a CME means a radio phenomenon figures out (partly or completely) the body of a CME.
- An indirect hint for a CME is given if preexisting radio sources are influenced / activated / associated with additional effects in a typical manner.

3.1   Direct Methods

Two approaches are possible:

- Detection of the CME by observing the radio counterpart of the features visible in coronograph images in its thermal radio emission.
- Detection of nonthermal radio emission associated with parts of the CME.

CMEs in Thermal Radio Emission. The detection of a CME as near as possible to its origin by imaging its incoherent radio emission (thermal free-free or gyroresonance emission) seems to be a difficult task due to the fact that the contrast between the CME and the background corona is not very strong at least during the beginning of its rise. The best choice might be difference images revealing the CME due to the movement of some structural features. Only two case studies have been reported in the literature (Sheridan et al. 1978, Gopalswamy and Kundu 1992, 1993). Sheridan et al. (1978) have seen a bulge in the quiet sun map at 80 MHz at the site of a simultaneously appearing white light transient and estimated a fourfold density excess from the radio data. Gopalswamy and Kundu (1992) found a laterally archlike extended enhancement at frequencies between 80 and 50 MHz resembling the optical CME feature which earlier passes the plasma levels observed by the imaging instrument. In this case, the radio source was not moving. Gopalswamy et al. (1996) traced the thermal microwave emission of a slowly erupting prominence which trails a "frontal loop - cavity" - CME pattern seen in YOHKOH SXT images.

Gopalswamy and Kundu (1992) give a CME mass estimate basing on the thermal bremsstrahlung process. Of course this method suffers from the same geometrical (unknown) parameters as mass estimates from white light images. The main problem is the superposition of CME- or flare related nonthermal radio sources which are stronger and hide the fainter thermal emission for the sensitive imaging instruments. On the other hand just these stronger nonthermal components are mostly the only sign of a CME in full disk patrol radio observations. In the following some attention is paid to these effects.

CMEs Traced by Nonthermal Radio Emission. A direct method to see CMEs in radio is tracing a nonthermally radio emitting component which can be associated with a CME (e.g. Hildner et al. 1986). Nonthermal radio emission evidences the presence of nonthermal electrons moving in or along a preexisting coronal (magnetic field) structure. Of course, nonthermal radio emission is also observed during solar flares. With regard to our initial remark (Section 2.1 and Figure 1a) about the different spatial scales of flares and average CMEs one must be careful with directly claiming evidence for having seen a CME without an independent confirmation.

A typical example is tracing plasma or gyrosynchrotron emission of the archlike or plasmoidlike moving type IV burst source (e.g. Stewart et al. 1982, Gopalswamy and Kundu 1989, Klein and Mouradian 1991). The moving type IV burst can sometimes be associated with the densest features visible in white light transients. However as with type II bursts the relationship between moving type IV sources and CMEs is still very unclear (Pick and Trottet 1988, Klein 1995). Aurass and Kliem (1992) draw the attention to a certain kind of type IV bursts with a characteristic late decimetric continuum pulse including a lot of spectral fine structures. They relate this observation to the disruption of the current flowing through the disconnected rising filament.

There are some reports about nonthermal phenomena which can be understood as the radio illumination of a CME by flare accelerated electrons. An example was given by Aurass et al. (1986), see Figure 8. With a time delay of 40 min, at two observing frequencies - 113 and 64 MHz - appears a morphologically highly similar pattern on the flux records. The corresponding

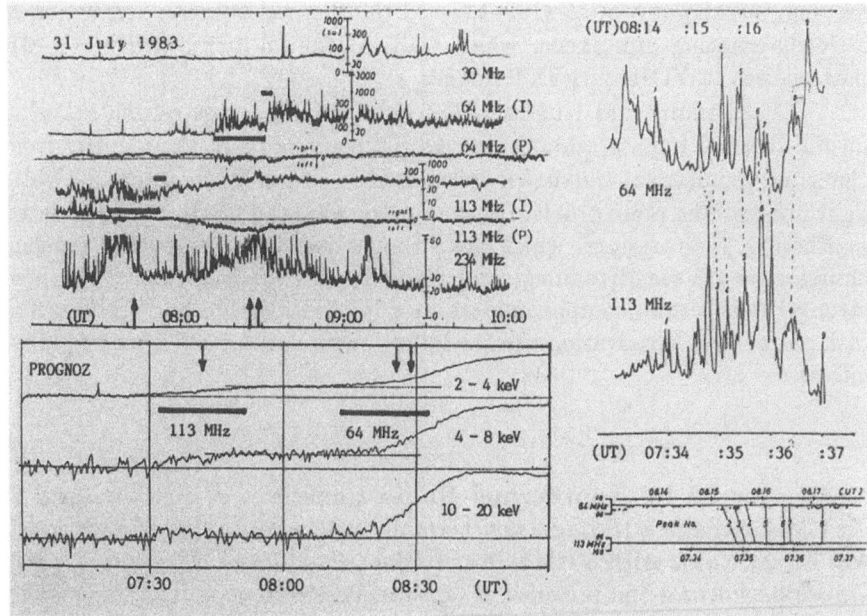

Fig. 8. Repetitive nonthermal radio bursts revealing the movement and the expansion of a long-living coronal structure on July 31, 1983 (Aurass et al. 1986).
Top left: the flux records, bars mark reappering intervals at 113 resp. 64 MHz. Arrows give times of subflares in the related active region. Right: High time resolution records of associated parts of both intervals, note the 40 min time delay and the time scale expansion. Bottom right: the spectral records of Weissenau confirm the narrow bandwidth of the effect. Bottom left: The soft X-ray records (PROGNOZ).

plasma levels have a radial distance of about 0.3 $R_\odot$. Comparing high time resolution records of these narrow band patterns at both frequencies reveals a time scale expansion between the observation at 113 and 64 MHz. A volume expansion rate of 0.2 - 1.2 % per minute has been estimated. The tail of the pattern expands faster than the head. The expansion is of the same order as for the giant X-ray arches analysed by Švestka et al. (1995), see Section 2.1. Because of the type I burst-like bandwidth of the patterns plasma emission is the relevant mechanism. The radio emission could be excited by nonthermal electrons enhancing plasma wave turbulence which is scattered directly into transversal waves at the clumpy, inhomogeneous medium represented by the CME (Melrose 1980). A confirmation of our interpretation was given by PROGNOZ soft X-ray data. During the appearance time of the flux patterns at the two different plasma levels subflares were reported. They are associated with faint soft X-ray enhancements. This is in some way similar to the results of Lantos et al. (1981) who showed a clear temporal association between narrow band coronal plasma emission and an soft X-ray burst. In our case, between the first and the second "illumination" the coronal structure moves outward. A speed of 90 km s^{-1} was estimated well within the range of CME velocities. What is conservatively and cautiously called "strange reappearing noise storm chains" in Aurass et al. (1986) was probably a flare induced radio illumination of parts of a CME.

3.2   Indirect Methods

Here we investigate two possibilities:

- A characteristic change of a preexisting noise storm continuum possibly accompanied or followed by a gradual microwave burst.
- A repeated appearance of groups of type III and reverse drift bursts with spatially distributed but simultaneously acting sources.

Changes of Preexisting Noise Storm Sources. Noise storms (Elgarøy 1977) consist of long duration (hours to days) enhanced and smoothly varying meter wave (about 50 - 500 MHz) continuum emission. Superposed to the continuum component are the type I bursts, an ensemble of short time and narrow band emission pulses. Noise storms reveal a long duration nonthermal energy release outside flares (Klein 1995) above magnetically complex active regions. They have different starting frequencies and overall bandwidth if their onset is flare- or non-flare-associated (Böhme 1993). Due to the long duration of noise storms and their frequency range - which corresponds to the height range in which CMEs are initiated - a disturbance of a noise storm by an (independent) CME is not improbable. There are reports in the literature about diffuse brightness enhancements in white light associated with noise storms source regions (Kerdraon and Mercier 1982, Kerdraon et al. 1983). From case studies it is known that noise storms can be stimulated, but also suppressed by flares and by CMEs.

Aurass et al. (1993, and references about flare related effects therein) have shown that chromospheric matter evaporated during a flare was rising with about 300 km s^{-1} and has suppressed a preflare noise storm continuum. The resulting radio spectrum shows a type II burst in absorption. Chertok (1993) and Kahler et al. (1995) point the attention on a typical sequence of events in association with CMEs:

- A preexisting noise storm starts to grow. The growth has a duration of about 10 minutes and appears with a negative drift rate corresponding with an apparent speed of some tens of km s^{-1}.

- The noise storm significantly decays within some minutes over a broad frequency range.
- In the same time interval a microwave burst starts to grow, sometimes with significant onset time delay from low to high frequencies. The microwave burst has no impulsive component, is not simply correlated with a soft X-ray burst and is sometimes referred to as "post eruptive energy release" (e.g. Chertok 1993) or belonging to "gradual events associated with CMEs" (e.g. Klein 1995 and references therein).

The smooth and long lasting meter wave continuum changes are difficult to recognize in routine spectral records. They are much easier to detect with a grid of single frequency receivers.

In the example Figure 9 (which we can not discuss in detail, here) the above mentioned phases are visible in the following intervals:

- The noise storm growth resp. the superposition of a slowly negatively drifting new radio source between 500 and 100 MHz starting at about 10:30 UT.
- The noise storm cessation starting some minutes past 11 UT and best visible between 300 and 200 MHz and from 11:20 till 11:30 UT.
- The gradual decimeter - microwave burst starting between 11:08 and 11:14 UT in the frequency range 500 MHz till 35 GHz (A. Magun, pers. comm.).

In the CME / radio literature the period February 1986 is of special interest due to the simultaneous action of the SMM coronograph and the Clark Lake Meter Wave Radioheliograph in an essential period of solar activity (reports on 13 - 17 February 1986 e.g. in Gopalswamy and Kundu 1992, 1993, Smith et al. 1993, Gopalswamy et al. 1994, Hundhausen 1994 in his Figures 31 and 32). We present in Figure 9 an example for a switch-off of a noise storm situated above the north-west limb (NRH 169 MHz) by a large scale disturbance. We argue that this disturbance is related with a CME. Burkepile and St.Cyr (1993) report on late hints for such a CME which was not directly observed due to SMMs schedule. Unfortunately, the analysis of several events of the February 1986 activity complex is scattered in literature - a comprehensive homogeneous analysis of all available data of this period seems promising.

Reverse Drift Bursts Can Indicate Coronal Destabilization. Klein and Aurass (1993) and Klein et al. (1996) describe the destabilization of the corona above a complex of four active regions. The loss of stability has been announced by preceding groups of meter wave fast drift bursts with a characteristic source distribution in space. Three 1.5 min duration groups of drift bursts were observed between 100 and 170 MHz during 3.5 hours before a filament eruption (Figure 10, left for the spectra). The burst groups consist

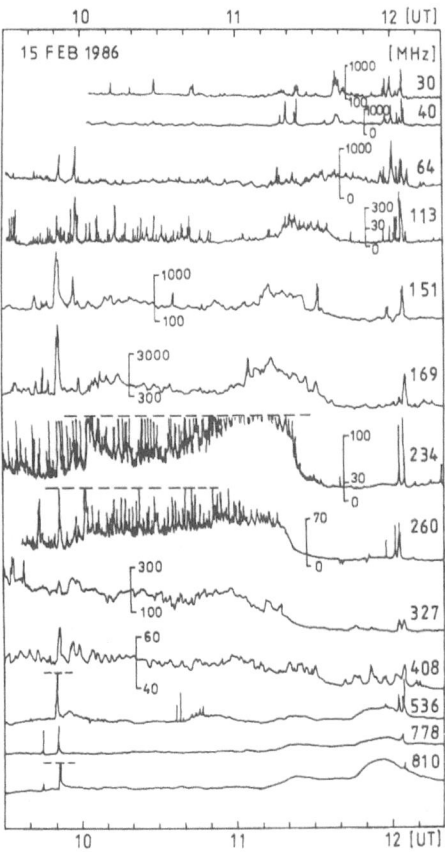

Fig. 9. The CME related noise storm enhancement, cessation and gradual microwave burst on February 15, 1986 around AR NOAA 4713 at the west limb. Single frequency records of Potsdam - Tremsdorf, NRH and AO Ondřejov, scales in solar flux units.

of type III bursts starting at frequencies larger than 165 MHz and of reverse drift (RS) bursts starting at about 120 MHz. We have drawn the RS burst sources as continuous crosses, the type III burst sources as stippled crosses over a composition of a Meudon H_α and a Mauna Loa coronograph image from the day before (from Klein et al. 1996). There are no white light CME observations available for the given day. As an inlay of Figure 10 right we show the involved active regions (among them two young regions displaying emerging flux).

Each of the spectrally simple RS bursts consists of two simultaneously brightening, widely displaced radio sources situated about 1 $R_\odot$ apart on both sides of a streamer-like configuration. This means the acceleration

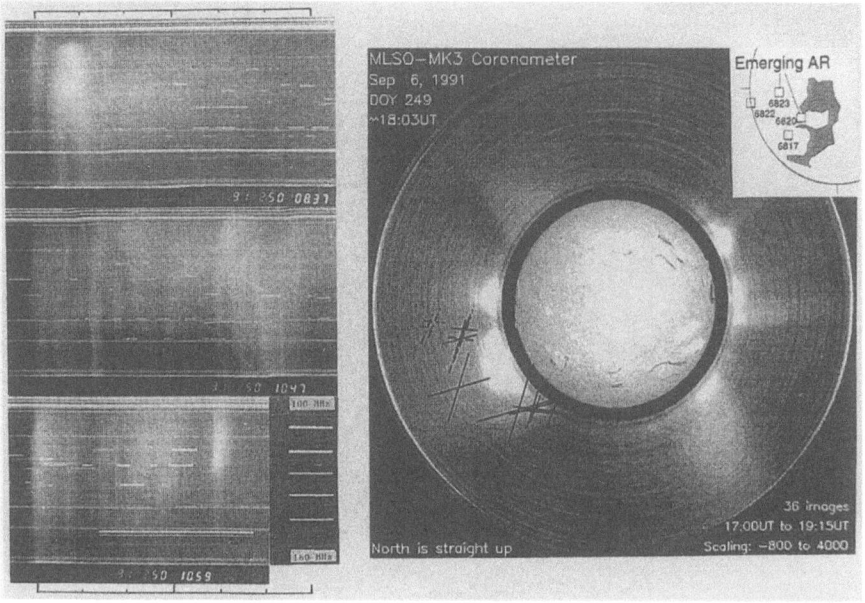

Fig. 10. Fast drift bursts in a destabilized part of the corona. Left : spectra 100 -
170 MHz, three time intervals from top to bottom, the bar denotes a two minute
scale. Right: coronographic and H_α data; radio sources as crosses superposed. For
details see text and Klein et al. (1996).

site of those electrons which become visible as RS drift burst is positioned
at heights > 1 $R_\odot$ above the photosphere. In contrast, the type III burst
sources are single ones and situated at the northern side of the streamer,
only. The type III beam injection takes place at a nearly simultaneously acting acceleration site below 0.5 $R_\odot$. A filament eruption in NOAA AR 6817
past 12 UT, accompanied by a moving type IV burst, is the final signature of
the destabilization of this part of the corona. During the observed process of
reaction of the corona to continuous energy supply, small scale energy release
processes were acting simultaneously in different heights of the involved part
of the corona. Probably Klein et al. (1996) have seen what Sime (1989) described: "... CMEs represent rapid evolution of previously formed magnetic
field structures through the gradual arrival of those structures at a state of
instability which leads to rapid dynamical evolution."

4   Summary

We have tried to summarize the current knowledge about common aspects of white light CMEs and solar radio type II bursts. Several open questions have been assembled in this review:

- What is the large scale type II burst source pattern in relation with the CME body ?
- Can we confirm radio emission from quasi-transversal and quasi-parallel shock transitions ?
- Are coronal shocks driven by flare blast waves, by flare ejecta, or by a CME piston ?
- Are there relations between the spectral fine structure of type II bursts and the three aforementioned aspects ?
- Can we successfully revise the CME / type II speed / shock directivity problem ?
- What is the relationship between the time scales of nonthermal (radio) and thermal (soft X-ray) energy release during flare associated CMEs ?
- Are noise storm cessation events associated with gradual microwave bursts, and reverse drift bursts useful CME predictors ?
- Are further nonthermal radio phenomena (beneath the moving type IV bursts) associated with the process of magnetic disconnection of filaments, plasmoids and CMEs ? What happens during the onset and the decay of the moving type IV burst ?

The combined analysis of decimeter and meter wave radio observations from ground with the SOHO LASCO coronograph and YOHKOH SXT images can solve some of these problems.

Acknowledgements: The paper was improved by the patient and constructive criticism of an unknown referee and of the editor, Dr. G. Trottet. The author gratefully acknowledges discussions with the late Dr. H.- W. Urbarz, as well as with Dr.s J. Burkepile, H.-T. Claßen, E. Cliver, A. Klassen, K.- L. Klein, G. Mann, N. Vilmer and E.Ya. Zlotnik. Further, he is grateful to M. Karlický, A. Magun and P. Zlobec for data supply. He thanks H. Detlefs, D. Scholz and B. Schewe for software support and help in preparing the figures, and the staff of Potsdam-Tremsdorf and Nançay Observatories for running the observations. The work was supported by the YOHKOH community with the generous access to the SXT data. The cooperation between the Paris-Meudon Observatory and the AIP was possible due to the travel grants 312/pro-bmft-gg (DAAD) and 94053 (MAE) within the German-French PROCOPE program. The visit of the author at the CESRA 1996 was supported by the Deutsche Forschungsgemeinschaft grant No. DFG/Au 106/6-1. Thanks are due to the organizers of this conference.

References

Aurass H. (1992): Annales Geophysica **10**, 359

Aurass H., Hofmann A., Kurths J., Mann G. (1986): Solar Phys. **107**, 129

Aurass H., Hofmann A., Magun A., Soru-Escaut I., Zlobec P. (1993): Solar Phys. **145**, 151

Aurass H., Klein K.-L., Mann G. (1994a): ESA-SP **373**, 95

Aurass H., Magun A., Mann G. (1994b): Space Sci. Rev. **68**, 211

Aurass H., Kliem B. (1992): Solar Phys. **141**, 371

Aurass H., Rendtel J. (1989): Solar Phys. **122**, 381

Averianikhina Y.A., Paupere M., Eliass M., Ozolins G. (1990): Astron.Nachr. **311**, 6, 367

Benz A.O., Thejappa G. (1988): A&A **202**, 267

Böhme A. (1993): Solar Phys. **143**, 151

Brückner G.E., Howard R.A., Koomen M.J., and 12 coauthors (1995): Solar Phys. **162**, 357

Bougeret J.L. (1985): in Tsurutani B.T. and Stone R.G. (eds.) "Collisionless Shocks in the Heliosphere", Rev. of Current Research, AGU GN-34, Washington D.C., 13

Burkepile J.T., St. Cyr O.C. (1993): NCAR/TN 369+STR, HAO, Boulder, Colorado

Cane H.V., Sheeley N.R., Jr., Howard R.A. (1987): JGR **92**, A9, 9869

Chertok I.M. (1993): Astron. Rep. **90**, 135

Pearson D.H., Nelson R., Kojian G., Seal J. (1989), ApJ **336**, 1050

Dryer M. (1994): Space Sci. Rev. **67**, 363

Dulk G.A. (1982): in Proc. CESRA Conf., Trieste, Italy, p. 356

Elgarøy Ø. (1977): Solar Noise Storms, Pergamon Press, Oxford

Gary D.E., Dulk G.A., House L.L., Illing R.M.E., Wagner W.J., McLean D.J. (1985): A&A **152**, 42

Gergely T.E., Kundu M.R., Erskine F.T., III, and 11 coauthors (1984): Solar Phys. **90**, 161.

Gopalswamy N., Hanaoka Y., Kundu M.R., Enome S., Lemen J.R., Akioka M., Lara A. (1996): APJ, submitted

Gopalswamy N., Kundu M.R. (1989): Solar Phys. **122**, 145

Gopalswamy N., Kundu M.R. (1992a): in Zank, G.P. and Gaisser, T.K. (eds.) Part. Acc. in Cosm. Plasmas, AIP Conf. Proc. **264**, 257

Gopalswamy N., Kundu M.R. (1992): APJ **390**, L37

Gopalswamy N., Kundu M.R. (1993): Solar Phys. **143**, 327

Gopalswamy N., Kundu M.R., St. Cyr O.C. (1994): APJ **424**, L135

Harrison R.A. (1991): Phil.Trans.Roy.Soc. London A**336**, 401

Harrison R.A., Hildner E., Hundhausen A.J., Sime D. (1990): JGR **95**, 917

Hiei E., Hundhausen A., Sime D. (1993): Geophys. Res. Letters **20**, 2785

Hildner E., Bassi J., Bougeret J.L., and 18 coauthors (1986): in Kundu M.R., Woodgate B.E. (eds.) Energetic Phenomena on the Sun, NASA CP-2439

Holman G.D., Pesses M.E. (1983): ApJ **267**, 837

Hudson H. (1995): http://isasxa.solar.isas.ac.ip/hudson/scratch/iau153paper.ps

Hundhausen A.J. (1993): Journ.Geophys.Res. **98**, A8, 13.177

Hundhausen A.J., Burkepile J.T., St. Cyr, O.C. (1994): JGR **99**, A4, 6543

Hundhausen A.J. (1995): Proc. High Energy Physics Conference, in press

Hundhausen A.J. (1994): in "The Many Faces of the Sun", to appear

Kahler S.W. (1991): ApJ **378**, 398

Kahler S.W. (1992): Ann.Rev.Astron.Astrophys. **30**, 113

Kahler S.W., Cliver E.W., Chertok I.M. (1994): in V. Rušin, P. Heinzl and J.-C. Vial (eds.), Solar Coronal Structures, VEDA Publ. Comp. Bratislava, p. 271

Kahler S.W., Sheeley N.R. Jr., Liggett M. (1989): ApJ **344**, 1026

Kahler S.W., Sheeley N.R. Jr., Howard R.A., Koomen M.J., Michels D.J. (1984): Solar Phys. **93**, 133

Kai K., Nakajima H., Kosugi T. (1983): PASJ **35**, 285

Karlický M., Odstrčil D. (1994): Solar Phys. **155**, 171

Klassen A. (1996): Solar Phys. **167**, 449

Klein K.-L. (1995): in Benz A.O., Krüger A. (eds.): Coronal Magnetic Energy Energy Release, Springer, Berlin, p. 55

Klein K.-L., Aurass H. (1993): Adv.Sp.Res. **13**, 9, 295

Klein K.-L., Trottet G., Benz A.O., Kane S.R. (1988): ESA SP-**285**, 157

Klein K.-L., Trottet G., Aurass H., Magun A., Michou Y. (1995): Adv.Space Res. **17**, 425, 247

Klein, K.-L. Aurass, H., Soru-Escaut I., Kalman B. (1996): A&A accepted

Klein K.- L., Mouradian Z. (1991): in Schmieder B. and Priest E. (eds.), Flares 22 Workshop on Dynamics of Solar Flares, Chantilly, France, p. 185

Kerdraon A., Mercier C. (1982): in Proc. CESRA Conf. Trieste, Benz A.O., Zlobec P. (eds.), 27

Kerdraon A., Pick M., Trottet G., Sawyer C., Illing R., Wagner W., House L. (1983): APJ **265**, L19

Kopp R.A., Pneuman G.W. (1976): Solar Phys. **50**, 85

Krall N.A., Trivelpiece A.W. (1973): Principles of Plasma Physics, McGraw-Hill, New York

Kundu M.R., Erickson W.C., Gergely T.E., Mahoney M.J., Turner P.J. (1983): Solar Phys. **83**, 385.

Kundu M.R., Gopalswamy N. (1992): in Švestka, Z., Jackson, B.V., Machado, M.E. (eds.), Eruptive Solar Flares, Springer, p. 268

Labrum N.R. (1985): in Dulk G.A., McLean (eds.), Solar Radio Physics, Cambridge Univ. Press, Cambridge, p. 155

Lantos P., Kerdraon A., Rapley G.G., Bentley R.D. (1981): A&A **101**, 33

Low B.C. (1993): Adv. Space Res. **13**(9), 63

Mann G. (1995a): in Benz A.O., and Krüger A. (eds.), Coronal Magnetic Energy Energy Release, Springer, Berlin, p. 183

Mann G. (1995b): J. Plasma Phys. **53**, I, 109

Mann G., Aurass H., Paschke J., Voigt W. (1992): ESA-Journal P-**348**, 129

Mann G., Claßen H.T., Aurass H. (1995): A&A **295**, 775

Mann G., Claßen H.T. (1995): A&A **304**, 576

Mann G., Klassen A., Claßen H.T., Aurass H., Scholz D., MacDowall R.J., Stone R.G. (1996): AIP preprint **95-26**, A&AS in press

Mann G., Klose S. (1995): in AGU 1995 Fall Meeting, Suppl. Eos **7**, F455

Mann G., Lühr H., Baumjohann W. (1994): JGR **99**, A13, 13315

McAllister A.H., Dryer M., McIntosh P., Singer H., and Weiss L. (1994): in Proc. of the Third SOHO Workshop, ESA SP-**373**, 315

McAllister A.H., Dryer M., McIntosh P., Singer H., and Weiss L. (1996): JGR **101**, A6, 13497

Melrose D.B. (1980): Space Sci. Rev. **26**, 3

Mouradian Z., Soru - Escaut I., Pojoga S. (1996): Solar Phys. **158**, 269

Nelson G.J., Melrose D.B. (1985): in Dulk G.A., McLean D.J. (eds.), Solar Radio Physics, Cambridge Univ. Press, Cambridge, p. 333

Pick M., Trottet G. (1988): Adv. Space Res. **8**, 11, 21

Robinson R.D. (1985): Solar Phys. **95**, 343

Sheeley N.R., Stewart R.T., Robinson R.D., Howard R.A., Koomen M.J., Michels D.J. (1984): ApJ **279**, 839

Sheridan K.V., Jackson B.V., McLean D,J., Dulk G.A. (1978): PASA **3**, 249

Sime D.G. (1989): JGR **94**, A1, 151

Smith K..L., Švestka Ž., Strong K., McCabe M.K. (1993): Solar Phys. **149**, 363

Steinolfson R. (1992): Proc. 26th ESLAB Symp., ESA SP-**346**, 51

Stewart R.T., Dulk G.A., Sheridan K.V., House L.L., Wagner W.J., Sawyer C., Illing R. (1982): A&A **116**, 217

Švestka Ž. (1976): Solar Flares, Reidel, Dordrecht.

Švestka Ž., Farnik F., Hudson H., Uchida Y., Hick P., Lemen J.R. (1995): Solar Phys. **161**, 331

The Radioheliograph Group (1993): Adv.Space Res. **13**, 9, 411

Tsuneta T. (1996): ApJ **456**, 840

Tsuneta S., Acton L., Bruner M., and 10 coauthors (1991): Solar Phys. **136**, 37

Uchida T. (1974): Solar Phys. **39**, 431

Vršnak B., Ruždjak V., Zlobec P., Aurass H. (1995): Solar Phys. **158**, 331

Vilmer N., Trottet G., Barat C., and 5 coauthors (1993): in Advances in Solar Physics, ed. by G. Belvedere, M. Rodonò, G. Simnett, Lecture Notes in Physics **432**, 197.

Wagner W.J., MacQueen R.M. (1983): A&A **120**, 136

Webb D.F. (1994): in J. Bergeron (ed.), Reports on Astronomy, **XXIIA**, p. 64

Webb D.F., Forbes T.G., Aurass H., Chen J., Martens P.C.H., Rompolt B., Rušin V., Martin S.F. (1994): Solar Phys. **153**, 73

Webb A.H., Howard R.A. (1994): JGR **99**, A9, 4201

Weiss A.A. (1963): Australian J. Phys. **16**, 240

Shock Waves and Coronal Mass Ejections*

Karl-Ludwig Klein[1] and Gottfried Mann[2]

[1] Observatoire de Paris, Section de Meudon, DASOP & CNRS-URA 2080,
   F-92195 Meudon Principal Cedex, France (klein@obspm.fr)
[2] Astrophysikalisches Institut Potsdam, Observatorium für solare Radioastronomie,
   Telegrafenberg A31, D-14473 Potsdam, Germany (gmann@aip.de)

Abstract. Shock waves and coronal mass ejections (CMEs) are large scale phenomena of solar activity. They are generated in the solar corona and can enter into the interplanetary space. Both shock waves and CMEs are accompanied by energetic particles (electrons, protons, and heavy ions). The paper summarizes working group discussions on the propagation of CMEs and shock waves in the corona, on the role of CMEs in energy release processes, and on microphysical processes creating enhanced plasma wave turbulence and particle acceleration at shock waves. The ULYSSES, WIND, and SOHO missions together with ground based observations and additional theoretical investigations are an exceptional opportunity to achieve progress in answering the many open questions.

Résumé. Ondes de choc et éjections de matière coronale (EMC) font partie des phénomènes à grande échelle spatiale de l'activité solaire. Générées dans la couronne, elles peuvent pénétrer dans l'espace interplanétaire. Les ondes de choc aussi bien que les EMC sont accompagnées de particules énergétiques (électrons, protons, ions lourds). On résume les discussions du groupe de travail sur la propagation des EMC et des ondes de choc dans la couronne, le rôle des EMC dans des processus de libération d'énergie, ainsi que sur des processus microphysiques qui génèrent des niveaux élevés de turbulence et des populations de particules énergétiques dans les ondes de choc. La combinaison des observations spatiales obtenues par ULYSSES, WIND et SOHO, avec des observations effectuées depuis le sol et des études théoriques nouvelles, offre une occasion exceptionnelle de mieux comprendre la physique de ces phénomènes.

1 Introduction

In the solar corona shock waves may be generated by flares and/or coronal mass ejections (CMEs). Flares represent a sudden energy release, in which stored magnetic energy is converted into mass motions, heating of the plasma and kinetic energy of accelerated particles. This energy release is accompanied with a large perturbation of the surrounding plasma, i.e., a large amplitude magnetohydrodynamic wave. Such a large amplitude MHD wave (blast wave) steepens into a shock wave according to its nonlinear evolution (cf. Priest 1982). On the other hand, CMEs represent large scale motions of dense and

---

* Working Group Report

bounded coronal material from the solar corona into the interplanetary space. They are mainly observed by extraterrestrial coronographs since Skylab (e.g., MacQueen 1980, Howard et al.1985, Munro and Sime 1985, and Hundhausen 1987). They have a typical spatial size of a solar radius in the corona. If the velocity of a rising CME exceeds the local Alfvén speed, a bow shock will be established ahead of the CME.

Shock waves generated by blast waves or CMEs manifest themselves in solar type II radio bursts (Wild et al. 1959, Smerd et al. 1962; cf. also Bougeret 1985, Nelson and Melrose 1985, Aurass 1992, and Mann 1995).

Both shock waves and CMEs are large scale disturbances in the solar corona and the interplanetary space. They are special forms of solar activity and need not be related to a solar flare observed by H_α or X-ray patrol instruments. Although these phenomena were investigated by different manners, i.e., observational and theoretical, since many years ago, a lot of questions are still open. Stimulated by the ULYSSES, WIND, and SOHO missions, solar and heliospheric physics will experience a great progress in the next decade. The aim of the working group *Shock Waves and Coronal Mass Ejections* was to discuss the present knowledge on the physics related to this subject, with emphasis on how radio diagnostics can be used jointly with the measurements from SOHO. The following text attempts to summarize the work of 2.5 half-day sessions and its scientific context. Much of the text reflects the authors' personal view. It is not understood as a review of present knowledge (cf. review by Aurass, *this volume*), but to fix some ideas that we hope will be further discussed in coming years during the joint exploration of SOHO and radio diagnostics of dynamical processes in the corona.

The discussions of the working group were divided into three parts:

- Coronal mass ejections: origin and consequences for energy release and particle acceleration
- Solar and interplanetary type II radio bursts
- Microphysical processes at shock waves in the corona and the interplanetary space

Since time was lacking, many subjects could only be alluded to. We collect a list of open questions in the final Section.

2   Coronal Mass Ejections and Large Scale Structures

There are several manifestations of the re-configuration of the corona which lead to the upward propagation of material. The eruption of active and quiescent prominences seen in H_α is the longest known evidence, but since the attained velocities are in general below the speed of gravitational escape, prominences are not qualified as "mass ejections". Moving type IV radio bursts attain sufficiently high velocities. However, since they are due to non-thermal electrons, their relationship with the coronal plasma as a whole is not clear, and they are generally considered as a peculiar feature with uncertain relevance to CMEs. So definition of CMEs and speculations about their magnetic structure comes from white-light coronographic observations. The complexity of transient features seen in these observations has been classified in different ways by different authors (e.g. Howard et al.1985, Burkepile and St. Cyr 1993) The example often presented as archetypal is the expanding loop-type configuration with or without embedded prominence (cf. discussion by Aurass, *this volume*). The main topics of the working group were the structure of CMEs and their relationship with energy release processes.

2.1   Structure of CMEs

The working group discussed radio signatures of CMEs in thermal bremsstrahlung, which might become a complementary diagnostic of the density distribution in the ejected structure (cf. Aurass, *this volume*). Soft X-ray images from YOHKOH have also been used to show large-scale transient structures, but were never before reported to trace the outer "loop" which is prominent in white-light images. N. Gopalswamy (cf. Gopalswamy et al. 1996) presented work with YOHKOH-SXT and the Nobeyama Radioheliograph at 17 GHz. He showed that the soft X-ray and microwave images display the same structure as white-light coronographic images of a CME, although on a smaller spatial scale (15 heliographic degrees): the rising prominence (microwaves), which has its long axis along the line of sight, lies within a density depleted region which is bordered by a steadily rising loop-shaped structure. The density of the "loop" inferred from X-ray images remains constant during its expansion, which implies that matter is feeded to the structure. The authors estimate a mass gain by a factor 4 in the course of the event.

Although the ejected material is likely not a low-β plasma, it is often assumed (implicitely or explicitely) that the observed plasma structures trace magnetic field configurations (cf. Hundhausen 1995, Fig. 6). In the simplest 2D picture the magnetic field lines are assumed to bridge the magnetic inversion line like in a dipole. However, the real magnetic structure is unknown for the time being. The simple configuration has several problems. First, when the structure is opened into interplanetary space, magnetic flux will be accumulated there (e.g. MacQueen 1980) which must be disconnected subsequently in order to ensure the constancy of magnetic flux observed at 1 AU (Kahler and Hundhausen 1992, Webb and Cliver 1995). Second, it is well known that prominence magnetic fields are more complex, leading to untwisting motions during eruption (e.g. Kurokawa et al. 1987, Rompolt 1990, review by Démoulin and Vial 1992). Theoretical work (e.g. Hughes and Sibeck 1987, Gosling et al. 1995) also suggests that three-dimensional reconnection will create helicoidal structures which may gradually detach from the Sun. In all these cases the magnetic field configuration of the ejecta is more complex

than in the simple scenario outlined above. Observational indications exist
that this is indeed relevant to the large scale of white-light CMEs:

1. Flux-rope or "plasmoid" configurations have recently been argued to
   explain circular structures and outward curved arcs identified in SMM
   coronographic observations (Webb and Cliver 1995).
2. Moving type IV bursts, which were only rarely seen as expanding loop-
   like features (Smerd and Dulk 1971), had been discussed in terms of
   plasmoids, i.e. closed magnetic field structures. In some cases they have
   been identified with localized features in the white-light CME (e.g. Stewart et al. 1982, Hildner et al. 1986, Gopalswamy and Kundu 1989), while
   in others they appeared to be related with the ejected prominence (e.g.
   McLean 1973, review by MacQueen 1980, Klein and Mouradian 1990).
   Such structures are also consistent with the more recent flux rope scenario.

Given that magnetic field measurements in CMEs are out of reach in the
near future, combined X-ray, visible light and radio observations are the only
means to clarify the magnetic topology of ejected structures. One tracer of
magnetic structures is the confinement of suprathermal electrons and their
relationship with processes in the underlying active region. Hence the clearer
identification of the relationship of Moving Type IV sources with the white-
light CME will likely give new insight into the magnetic structure of CMEs
as a whole.

2.2   Origin

It has become clear that CMEs are not in general the consequence of the
sudden energy release occurring during a flare: on the one hand the most
frequently associated transient phenomena in the low atmosphere are prominence eruptions. On the other hand, *extrapolated* starting times of CMEs
(Harrison et al. 1990) have been shown to precede the associated flare in
well documented single events, and this has been corroborated with the help
of ground-based coronographic observations which operated at lower altitude than space-borne coronographs (Hundhausen 1987, 1995; but counter-
examples exist - see Dryer 1994). Since in the case of filament eruptions it
is found without ambiguity that the filament lifts off before the flare starts
(Kahler et al. 1988), it appears justified to make a similar statement for
CMEs.

   The impossibility to observe directly the start of a CME remains nonethe-
less a major uncertainty in establishing the sequence of events during magnetic energy release to bulk motion and other forms. Gopalswamy's X-ray
and microwave observations appear indeed to differ from the timing analysis of white-light coronographic data: They observe especially that the outer
"loop" started at about 0.3 R_S above the photosphere. This structure apparently existed already in the phase of pre-eruptive prominence activity and

started to move outward during the initial slow rise of the prominence. This is different from the rare cases where the presumed onset of a (white-light) CME could be observed: in those events the first manifestation seemed to be the rise of the low-density region, identified with the filament cavity (cf. discussion by Hundhausen (1987) and references therein). The formation of the outer "loop" followed. There is an evident interest to use the combined white-light and XUV observations from SOHO in comparative studies of this kind to probe the evolution of the different components of a mass ejection prior to and during the instability.

2.3   Interplanetary Manifestations

Interplanetary disturbances can be identified by an increase of density fluctuations in comparison to the surrounding solar wind. P.-K. Manoharan presented a study of scintillation observations obtained with the Ooty Radio Telescope. He analyzed a large sample of (nearly 770) interplanetary disturbances between the solar minimum in August 1986 and the following maximum April 1991. The transients are predominantly generated near the equatorial region of the Sun during the minimum period. As the solar activity increases they are observed at higher latitudes and, sometimes, in the polar region during the maximum phase of solar activity. The propagation speeds derived from the scintillation measurements are in the range 200–1500 km/s with a mean value at 450 km/s. This agrees with previous studies (cf. Nelson and Melrose 1985). The latitude distribution of the inferred coronal region where the perturbations originate is very similar to that of CMEs and streamers observed by Hundhausen (1993). This makes scintillation observations an important tool to trace the propagation of large-scale coronal disturbances into the interplanetary medium.

2.4   Energy Release Processes Related with CMEs

The question how the ejection of coronal magnetic structures affects the ambient atmosphere is intimately related to that of the magnetic structure of CMEs. A scenario has been developed for prominence eruptions where the rising prominence opens the formerly closed magnetic field lines into a configuration with oppositely directed open field lines which are separated by a vertical current sheet (e.g. Kopp and Pneuman 1976). Closed loop structures are subsequently formed by reconnection across this current sheet. They are supposed to be the seat of the soft X-ray emission following prominence eruptions. When field lines farther and farther away from the current sheet reconnect, higher and higher loops form. In a more realistic three-dimensional scenario (Hughes and Sibeck 1987, Gosling et al. 1995) the reconnection between opposite "legs" of neighbouring sheared loops leads to the formation of helical magnetic field structures in the corona with their extremities anchored in the photosphere, and closed field lines underneath. In the following we include this 3D model in what we call a "Kopp-and-Pneuman-type scenario".

Many observations of prominence eruptions cope with this description: H_α flare ribbons, which are thought to trace the chromospheric footpoints of the coronal loops, lie on either side of the filament channel and gradually recede from each other. Full-Sun imaging in X-rays with YOHKOH-SXT revealed the formation of arcades bridging the photospheric inversion line at the place where the filament or prominence had previously been located:

- Hiei et al. (1993) observed that at the place where a prominence and an overlying helmet streamer had disappeared, a new streamer structure forms whose outer parts are visible as a slowly rising, cusp shaped X-ray structure.
- McAllister et al. (1994) and Alexander et al. (1994) observed the formation of an arcade of X-ray loops spreading within several hours over a range of 150^0 in longitude to either side of an initial arch structure across the photospheric inversion line. The time profile of the X-ray flux has a slow rise and fall over more than one day (Hudson et al. 1995).
- Tsuneta (1996) analyzed the temperature distribution in a cusp-shaped X-ray structure which forms and rises over several hours near the limb. The temperature estimated from X-ray images with different filters shows that at each time during the decay phase of the event the outer loops are the hottest, as expected if these loops are the most recently formed. It should be noted, however, that the temperature structures is more complex during the rise phase of the event, since different patches of high temperature are identified near the top, in one foot and throughout the other leg of the loop. This is also the phase during which energetic particles are accelerated as revealed by their hard X-ray emission (Hudson et al. 1994).
- Hanaoka et al. (1994) used imaging observations in H_α, soft X-rays and at 17 GHz to trace the eruption of a quiescent prominence and the ensuing formation and rise of an arcade of loops. H_α ribbons are observed at the footpoints of the arcade. The time profiles of both the radio and soft X-ray fluxes are similar in different regions of the arcade, suggesting a similar evolution of the temperature throughout the structure.

Gopalswamy's report on his similar observations as those of Hanaoka et al. (1994) showed also that the X-ray flux was enhanced in a large region during several hours, resembling a "long decay event", although of much weaker flux than the events detectable by GOES patrol observations. The most impulsive increase was observed in a bright knot underneath the centroid of the rising prominence (recall that the long axis of the prominence is along the line of sight). As this is the region where reconnection separates the flux rope from the closed loop structures, Gopalswamy's observation is again a qualitative confirmation of the Kopp-and-Pneuman scenario. However, the observations reveal also that at the place where the prominence had been located prior to its rise, a new prominence forms. The prominence formation starts about 13 minutes before the presumed X-ray arcade brightens. This shows that closed field lines – those containing the new prominence – exist in the region after the prominence started to lift off and *before* the post-eruptive arcade forms. This timing deserves further study, since it seems to suggest that either closed magnetic field structures form before the post-eruptive arcade becomes visible – although the latter is generally considered as the signature of reconnection and formation of closed loops – or else that the magnetic field configuration was not completely opened by the rising prominence.

I. Chertok reported on his attempts of ordering various phenomena of energy release as a function of the presence or absence of a CME (cf. also Chertok 1996). He considers flares and CMEs as a priori independent phenomena. In this picture the important contribution of a CME is that it creates the possibility of post-eruptive energy release within the scenario outlined above. The character and intensity of the ensuing manifestations depend on the magnetic field configurations which are affected by the CME. These manifestations may be mainly thermal, if weak fields far from active regions are affected (e.g. in association with prominence eruptions), while prolonged relativistic particle acceleration as observed by CGRO (Kanbach et al. 1993) and GAMMA (Leikov et al. 1993, Akimov et al. 1994, 1996) may occur if the CME opens strong magnetic fields above active regions. Chertok presented IZMIRAN observations of the microwave flux densities and GOES soft X-ray observations in complement to Feynman and Hundhausen's (1994) comparative study of the March 1989 activity. The ensemble of this data shows more clearly than other published cases that among X-class soft X-ray events those which occur together with a CME have a prolonged intense emission after the impulsive phase.

Irrespective of the successes of Kopp-and-Pneuman-like models in explaining prominence eruptions and their consequences, it is clear that their application to CMEs implies a change of the spatial scale. While Hiei et al. (1993) and McAllister et al. (1994) did observe large-scale phenomena that fit the Kopp-and-Pneuman scenario, the symmetry of this picture is in contradiction with the events studied by Harrison et al. (1990) where the X-ray loops arising in the aftermath of a CME are located on one side of the white-light CME, not below its centre. It must be kept in mind that these observations (XRP-SMM) were carried out with a restricted field of view in X-rays such that the large-scale structures seen by Yohkoh-SXT could not be imaged. Nevertheless the spatial scales of (post-)flare loops and H_α ribbons are much smaller than white-light CMEs. Hence even during a CME the organization and dynamics of the magnetic field seems to play a role in energy storage and release on the spatial scale of one active region and less, rather than on the scale of the global white-light structure.

As an alternative to directly creating the conditions for energy release, a CME, as a large-scale coronal perturbation, might act to trigger instabilities and energy release at different sites of the atmosphere where the local conditions – e.g. plasma-magnetic field interactions in a given active region – had led to energy storage. K.-L. Klein presented joint H_α and decimetric/metric radio observations of activity at the time of a prominence eruption (cf. Klein and Mouradian 1990). The global timing is as inferred from the filament eruptions studied by Kahler et al. (1988). The prominence lift off is followed within a few minutes by type III emission and a flare. However, the flare is about one solar radius away from the prominence. Nevertheless, both activities do not seem to occur together simply by coincidence. Indeed, a connection between both sites is suggested by fast-drift bursts which trace a large-scale loop structure in the imaging observations of the Nançay Radioheliograph. Coronographic observations do not exist for this event. But the relation which might be suggested between the different manifestations of activity is that events in localized regions are triggered by a common large-scale disturbance. Available data do not provide a clear answer, but joined imaging observations at XUV, white light and radio wavelengths should clarify the situation.

3   Solar and Interplanetary Type II Radio Bursts

In dynamic radio spectra solar type II radio bursts appear as stripes of enhanced radio emission slowly drifting from high to low frequencies. Generally, the spectral features of solar type II bursts comprise two components, the "backbone" and the "herringbones". The "backbone" is the aforementioned emission stripe with a slow drift of roughly $-0.1\,\mathrm{MHz/s}$ (Nelson and Melrose 1985). The "herringbones" are rapidly drifting emission stripes shooting up from the "backbone" to high and low frequencies. Because of their high drift rate of about $\pm\,10\,\mathrm{MHz/s}$ (Cairns and Robinson 1987) they resemble type III radio bursts.

Generally, the radio radiation is assumed to be emitted near the local electron plasma frequency in the meter and decameter wave range, i.e., $f_{\mathrm{obs}} \approx \omega_{\mathrm{pe}}/2\pi$ with $\omega_{\mathrm{pe}} = (4\pi e^2 N/m_e)^{1/2}$ (e, elementary charge; N, particle number density; m_e, electron mass). Because of the density inhomogeneity in the solar corona and the interplanetary space the drift rate $D_f = df_{\mathrm{obs}}/dt$ is related to the velocity V of the radio source by

$$D_f = \frac{f_{\mathrm{obs}}}{2} \cdot \frac{1}{N} \cdot \frac{\mathrm{d}N}{\mathrm{d}s} \cdot V \tag{1}$$

Here, dN/ds denotes the gradient of the density along the propagation path of the radio source. Note that solar type II radio bursts are mainly observed below 150 MHz. In order to estimate the velocity associated with the "backbone" and the "herringbone" the plasma parameters usually found at the 70 MHz level are employed, i.e., a particle number density $N = 7 \cdot 10^7\,\mathrm{cm}^{-3}$, a temperature $T = 2 \cdot 10^6$ K, a density height scale $(N^{-1} \cdot dN/ds)^{-1} = 2 \cdot 10^5$ km and a magnetic field strength $B = 1$ G (Dulk and McLean 1978).

These parameters result in an Alfvén speed $v_A = 260$ km/s, a sound speed $c_s = 170$ km/s, and a thermal electron velocity $V_{the} = 5500$ km/s. Then, the velocity of the radio source associated with the "backbone" and the "herring-bone" are estimated to be 570 km/s and 57 000 km/s, respectively. Therefore, the "backbone" and the "herringbones" are interpreted as the radio signature of a shock wave travelling (with a super-Alfvénic velocity) through the solar corona and electron beams accelerated at the shock wave, respectively (Nelson and Melrose 1985).

N. Gopalswamy reported on solar type II radio bursts recorded by the HIRAISO spectrometer (spectral range 25–2000 MHz) and the WAVES instrument aboard the WIND spacecraft. The WAVES instrument (Bougeret et al. 1995) receives the radio radiation in the range 20 kHz–14 MHz. The study included all type II bursts discovered in HIRAISO during the WIND operations. The author did not find any signatures in the radio data received by the WAVES instrument. Since the 14 MHz level corresponds to a distance of $\approx 2.25 R_S$ from the Sun (R_S, solar radius) he concluded that the shock waves associated with the type II bursts are not able to reach the outer corona and the interplanetary space. This observational result is regarded by him as a serious hint for a shock generated by a blast wave, since local energy density in a blast wave is decreasing during its evolution. This study supported the result recently obtained by Gopalswamy and Kundu (1992). They investigated 12 events of solar type II bursts and CMEs. For these events positional informations simultaneously obtained by the SKYLAB, SOLWIND and SMM-C/P white light instruments and by the Culgoora and Clark Lake Radioheliograph showed that 8 solar type II bursts had a source location well below the leading edge of the associated CME. Basing on this study (Gopalswamy and Kundu 1992) and the events presented at this workshop N. Gopalswamy concluded that solar type II radio bursts are mainly generated by blast waves produced by flares, while interplanetary type II bursts are piston driven due to a transient CME. In contrast to the conclusion by Gopalswamy and Kundu (1992) the observational fact, that some type II burst sources lie below the leading edge of the CME, can also be explained in the following way in terms of the model by Holman and Pesses (19983): The CME is travelling outwards in the corona. Because of its super-Alfvenic speed it drives a shock wave in front of itselfs. At the flanks of this bow shock a nearly perpendicular shock geometry, i. e., $\theta_{n_s,B_{up}}$, is exsisting and, consequently, shock drift acceleration can efficiently act (cf. Sec.4).

S. Hoang and P. Zlobec presented the characteristics of several examples of low drift bursts observed at very low frequencies (1.25 kHz–48.5 MHz) by the URAP instrument (Stone et al. 1992) aboard the ULYSSES spacecraft. The URAP instrument is able to receive the interplanetary radio radiation in the frequency range 1.0–940 kHz. The considered events appeared in the period November 1990 – February 1991. Low drifting events are classified in "sticks" and "backbones" according to their instantaneous bandwidth. The "sticks" resemble the "herringbones" in solar type II radio bursts. The mean lifetime of both types of events is practically the same, i.e., roughly 5 hours. According to a heliospheric density model (Fainberg and Stone 1971) usually employed a radial velocity of the agent producing the "backbones" and the "sticks" can be derived from their drift rate. Thus, a speed of 600 km/s was found. Furthermore, the source of these features is localized to be at 1.4 AU. Their relative instantaneous bandwidths $\Delta f/f \approx 0.02$ are quite smaller than those of solar and interplanetary type II radio bursts above 30 kHz (Lengyel-Frey and Stone 1989). Lengyel-Frey and Stone (1989) reported on instantaneous bandwidths in the range $\Delta f/f = 0.5 - -1.0$ in the case of kilometric type II's. In the metric and decametric wave range a typical value $\Delta f/f \approx 0.32$ was found for type II bursts (Mann et al. 1994a). An association with inter-planetary shocks is revealed for several cases of these low drifting phenomena.

K.-L. Klein investigated a sample of solar type II radio bursts simultane-ously observed by the radiospectrometer of the ETH Zürich and the radiohe-liograph of Paris Observatory in Nançay. The radial source velocity V_{SG} de-duced from the drift rate in the dynamic radio spectra is significantly smaller than the source speed V_{HG} measured by the radioheliograph in all cases, i.e., $V_{SG} < V_{HG}$. But the study reveals the relationship $V_{SG} \approx V_{HG} \cdot cos(\theta_{n_s,n_r})$, where θ_{n_s,n_r} denotes the angle between the shock normal $\mathbf{n_s}$ and the radial normal $\mathbf{n_r}$. This angle is estimated from the heliographic data. This can be explained if type II emission comes from a restricted domain of the shock, as suggested by the measurement of source sizes. Then even if the nose of the shock propagates radially outward, that part of the shock which is outlined by radio emission does not. The reason may be preferential acceleration regions (cf. Holman and Pesses 1983) or inhomogeneities in the ambient medium (Wagner and MacQueen 1983). As a consequence, the shock speeds deduced from the drift rate of type II bursts might be generally underestimated with respect to the true shock speed.

H. Aurass presented observations of a multi-lane type II burst event with the Weissenau spectrometer (30-1000 MHz). Some of the lanes have a zero drift rate. These lanes are seen to be emitted by an azimuthally moving (presumably super Alfvenic) radio source in 164 MHz Nançay radio helio-graph data. Azimuthally propagating shocks are able to accelerate particles at places far away from the flare region as previously assumed by Wibberenz et al. (1989).

The special solar type II radio burst appeared on September 27, 1993 evi-dently showed a structure of the emission bands ("backbone") at frequencies related as $1 \div 2 \div 3$. The simultaneous observations of this radio burst by the radiospectrometer of the Astrophysikalisches Institut Potsdam in Tremsdorf and the radioheliograph of the Observatory Paris-Meudon in Nançay allow to determine the brightness temperatures of the radio emission in the three emission bands. In contradiction to the second harmonic emission, which is well understood as the result of the coalescence of two high frequency electrostatic plasma waves into radio waves, the generation of the third harmonic can be done by different mechanism. E. Ya. Zlotnik offered a new mechanism for the third harmonic emission by nonlinear coalescence of three high frequency electrostatic plasma waves into escaping electromagnetic (radio) waves. The probability of this interaction has been calculated with the aim to estimate the brightness temperature of the third harmonic radio emission. This mechanism is also compared with those previously discussed by Cairns (1987a-c). The comparison of these theoretical results with the special type II burst observation on September 27, 1993 shows that the proposed mechanism can be responsible for the third harmonic emission, if the phase velocity of the coalescing plasma waves is quite small and the turbulence level of theses waves is strongly enhanced.

4   Microphysical Processes at Shock Waves

Shock waves are discontinuities at which plasma is heated and particles are accelerated. Since shock waves are able to accelerate electrons up to sub-relativistic energies, these shock waves can mediately produce radio waves. The accelerated electrons are forming an unstable distribution function. Such an electron population is able to excite Langmuir waves and/or upper hybrid waves, which can convert into escaping electromagnetic (radio) waves by non-linear interaction with ion density fluctuations and/or low frequency plasma waves (Melrose 1985). Thus, electron acceleration at shock waves is a very important problem from a radioastronomical point of view.

Coronal shock waves can only be observed by remote sensing techniques, e.g., radioastronomical methods as aforementioned. On the other hand inter-planetary shock waves and Earth's bow shock can be studied by extraterres-trial in-situ measurements as special examples of collisionless shocks in space plasmas. If one assumes that the basic physical processes are essentially the same in all collisionless shocks, one can learn from the in-situ measurements of interplanetary shocks and Earth's bow shock for a better physical under-standing of shock waves in the corona.

A very important question is, where is the source of radio emission lo-cated at shock waves, i.e., in the upstream or downstream region or in the transition zone, since the radio source region is also the region of electron acceleration. In-situ measurements of plasma waves at interplanetary shocks evidently show that high frequency plasma waves (upper hybrid and Lang-muir waves) are predominantly observed in the upstream region, while the low frequency plasma waves (e.g., whistler waves and ion acoustic waves) are mostly observed in the downstream region (Gurnett et al. 1979, Kennel et al. 1982, Thejappa et al. 1995). Because of these observations Mann et al. (1994a) concluded that the source region of the "backbone" emission should be located in the vicinity of the transition zone, where both the high and low frequency plasma waves occur. Note that both the high and low frequency plasma waves are needed for the fundamental radio emission. On the other hand, Lengyel-Frey (1992) argued by means of studying 20 interplanetary type II radio bursts that the radio radiation is coming from the downstream region. Filbert and Kellog (1979), Treumann et al. (1986), and Cairns (1986) investigated the radio radiation at Earth's bow shock by extra-terrestrial measurements and demonstrated that the radio emission is generated in the upstream region of the quasi-perpendicular part of Earth's bow shock. This is also confirmed by in-situ measurements of electron distribution functions by the 3D plasma instrument (Lin et al. 1995) aboard the WIND spacecraft in front of Earth's bow shock.

Shock drift acceleration is very efficient at quasi-perpendicular shocks, A fast magnetosonic shock wave is accompanied with a compression of the density and the magnetic field. Thus, it represents a magnetic mirror at which particles can be reflected and accelerated (Sonnerup 1969, Paschmann et al. 1980, Schwartz et al. 1983, Krauss-Varban 1989). During such a reflection the particle receives a velocity gain due to the action of the electric field within the transion zone of the shock wave. If $V_{i,\parallel}$ is the initial velocity the particle is accelerated up to a velocity $V_{r,\parallel}$ given by

$$V_{r,\parallel} = 2v_s \cdot \sec(\theta_{n_s,B_{up}}) - V_{i,\parallel} \qquad (2)$$

after a single reflection at the shock wave. Both $V_{i,\parallel}$ and $V_{r,\parallel}$ are velocity components parallel to the upstream magnetic field in the rest frame of the upstream plasma. Here, v_s and $\theta_{n_s,B_{up}}$ denote the shock speed and the angle between the shock normal $\mathbf{n_s}$ and the upstream magnetic field $\mathbf{B_{up}}$, respectively. Holman and Pesses (1983) suggested that the suprathermal electrons required for the type II burst emission are generated by shock drift acceleration. But such electrons can only be produced by this mechanism if the angle $\theta_{n_s,B_{up}}$ exceeds $80°$, i.e., if the shock wave is nearly perpendicular.

Up to now, the behaviour of a single particle has been considered during the reflection at the shock wave. If a particle population with a Maxwellian distribution function

$$f_i(V_{i,\parallel}, V_{i,\perp}) = \frac{1}{(2\pi v_{th}^2)^{3/2}} \cdot \exp\{\frac{-(V_{i,\parallel}^2 + V_{i,\perp}^2)}{2v_{th}^2}\} \qquad (3)$$

(v_{th}, thermal speed) is existing as the inital state in the upstream region, a shifted loss-cone distribution

$$f_r(V_{r,\parallel}, V_{r,\perp}) = \frac{\Theta(V_{r,\parallel} - U_s)}{(2\pi v_{th}^s)^{3/2}} \cdot \Theta(V_{r,\perp} - [V_{r,\perp} - U_s] \cdot \tan \alpha_{lc})$$

$$\times \exp\{\frac{-[(-V_{r,\parallel} + 2U_s)^2 + v_{r,\perp}^2]}{2v_{th}^2}\} \qquad (4)$$

($U_s = v_s \sec \theta_{n_s,B_{up}}$) results for the reflected particles due to shock drift acceleration (Leroy and Mangeney 1984, Wu 1984). Here, Θ is the well-known step-function. $V_{r,\parallel}$ and $V_{r,\perp}$ denote the velocity component parallel and perpendicular to the upstream magnetic field of the reflected particles, respectively. The loss-cone angle α_{lc} is defined by $\alpha_{lc} = \arcsin(B_{up}/B_{down})^{1/2}$ (B_{down}, magnitude of the magnetic field in the downstream region). R. P. Lin reported on a special electron distribution recently measured by the 3D plasma instrument aboard the WIND spacecraft. This distribution function was observed near the magnetic field line tangentially touching the Earth's bow shock. At this region the Earth's bow shock has a nearly perpendicular shock geometry. He demonstrated that the measured distribution function agrees very well with that given by Eq. (4). This observational fact support the solar type II burst model originally proposed by Holman and Pesses (1983) and improved by Benz and Thejappa (1988). According to this model the electrons are accelerated at a nearly perpendicular shock wave in the solar corona. These suprathermal electrons are establishing a shifted loss-cone distribution, which is unstable and excites upper hybrid waves (Benz and Thejappa 1988). These upper hybrid waves convert into escaping radio waves by the mechanism already mentioned in this Section.

G. Mann presented a mechanism for accelerating electrons to suprathermal velocities at quasi-parallel shock waves under coronal circumstances. As well-known from extraterrestrial in-situ measurements a super-critical, quasi-parallel shock wave in collisionless plasmas is accompanied by large amplitude, low frequency magnetic field fluctuations in the upstream and downstream region (Kennel et al. 1985). Earth's bow shock is the most investigated collisionless shock in space plasmas. Recently, so-called SLAMS (Short Large Amplitude Magnetic Field Structures) have been observed as a common feature of the quasi-parallel region of Earth's bow shock (Schwartz et al. 1992). Schwartz and Burgess (1991) argued that a supercritical, quasi-parallel shock transition should be considered as a patchwork of SLAMS. SLAMS are characterized as well-defined single magnetic field structures with large amplitudes of ≈ 2 times or more the upstream magnetic field and short durations of typically 10 s (Schwartz et al. 1992, Mann et al. 1994b). They are propagating quasi-parallel to the upstream magnetic field with a super-Alfvénic velocity (Schwartz et al. 1992, Mann et al. 1994b). Note that the maximum magnetic field compression within SLAMS is stronger than the jump of the magnetic field across the shock according to the Rankine-Hugoniot relations, i.e., $(B_{max}/B_{up})_{SLAMS} > (B_{down}/B_{up})_{RH}$. Furthermore, the speed of the SLAMS is increasing with its magnetic field compression. Consequently, two neighbouring SLAMS with different magnetic field compresssions represent two converging magnetic mirrors. Such a configuration is able to accelerate electrons. It was shown that electrons can be accelerated by this mechanism from thermal energies up to subrelativisic energies under coronal circumstances (Mann and Claßen 1995). Thus, the accelerated electrons have typical velocities of 60000 km/s (≈ 10 keV). Such electrons may be responsible for the "herringbones" (Cairns and Robinson 1987). A detailed study shows that this mechanism is producing a so-called cone mantle shaped distribution function

$$f_{\rm r} = \frac{\Theta(V_\parallel)}{(2\pi v_{\rm th}^2)^{3/2}} \cdot \delta(V_{\rm r,\perp} - V_{\rm r,\parallel}\tan\alpha_{\rm lc,SLAMS}) \cdot \exp\{\frac{-[(V_{\rm r,\parallel}/\nu)^2 + V_{\rm r,\perp}^2]}{2v_{\rm th}^2}\}(5)$$

for the accelerated particles, where the initial distribution was a Maxwellian (Mann and Claßen 1995). The angle $\alpha_{\rm lc,SLAMS}$ is defined by $\alpha_{\rm r,SLAMS} = \arcsin(B_{\rm up}/B_{\rm max,SLAMS})^{1/2}$ This distribution function (cf. Eq. (5)) has a beam like (i.e., $\partial f_{\rm r}/\partial V_{\rm r,\parallel}$) and a loss-cone like (i.e., $\partial f_{\rm r}/\partial V_{\rm r,\perp}$) part. Therefore, it is unstable and gives rise to both upper hybrid waves and Langmuir waves (Marsch 1990). Thus, the cone mantle distribution differs from a pure beam distribution, which is responsible for generating type III radio bursts.

In contradiction to the mechanism presented by Mann and Claßen (1995), V. Krasnoselskikh favours the acceleration mechanism proposed by Holman and Pesses (1983), i.e., shock drift acceleration, which is efficiently acting at nearly perpendicular shock waves. He argued that a shock wave travelling through the corona should not be regarded as a plane discontinuity, but a as spatially structured one. If a shock wave propagating nearly perpendicular to the ambient magnetic field has a shock front with a concave curvature with respect to the shock normal $\mathbf{n}_s$, a part of the electrons can be trapped in front of the shock wave. Consequently, these electrons gain energy due to multiple encounters with the flanks of the concave shock front via shock drift acceleration. Within such a configuration electrons are rapidly acclerated up to relativistic energies.

Thus, there are different mechanisms explaining the production of supra-thermal electrons needed for the type II burst emission at coronal shock waves. The decision, which of them reflects the reality, will be provided by observations, in particular in-situ measurements of energetic electron spectra and plasma waves. Such observations were and are done at interplanetary shocks and Earth's bow shock. Gurnett et al. (1979) reported on plasma wave spectra measured by HELIOS 1 and 2 at an oblique ($\theta_{\rm n_s,B_{up}} = 47.5°$) interplanetary shock. High frequency electron plasma oscillations (i.e., upper hybrid and/or Langmuir waves) were predominantly appearing in the up-stream region and transition region of the shock, while low frequency plasma waves (i.e., ion-acoustic and whistler waves) were occuring in the transition region and downstream region of the shock waves. The same was qualitatively observed at several interplanetary shocks with $22° \le \theta_{\rm n_s,B_{up}} \le 88°$ by Kennel et al. (1982) using the data of the plasma wave instrument aboard ISEE 3. Recently, Thejappa et al. (1995) analyzed plasma wave spectra at five inter-planetary (reverse) shocks by evaluating the data of the URAP instrument aboard ULYSSES. At a quasi-parallel shock with $\theta_{\rm n_s,B_{up}} = 35°$, which oc-curred on January 22, 1993 (cf. Fig. 2 in the paper by Thejappa et al. (1995)) an enhanced plasma wave turbulence at the electron plasma frequency $f_{\rm pe}$ has been observed in the upstream region and the transition region. This turbulence disappeared in the downstream region. These observational facts show that an enhanced level of electrostatic electron oscillations appears in the upstream region of both quasi-parallel and quasi-perpendicular shocks. This indicates that electrons must be accelerated by both types of shocks. But a further investigation of this subject is necessary in order to answer the question: Which mechanism accelerates the electrons up to suprathermal and relativistic velocities at shock waves in the solar corona and the interplanetary space ? Here, the participants of the working group expect new results from the data received by ULYSSES, WIND and SOHO with respect to this subject.

5   Final Remarks and Open Problems

Coronal mass ejections and large-scale shock waves are topics where remote sensing at radio wavelengths significantly contributes to our understanding of basic processes in cosmic plasmas. The workshop discussions demonstrate that the presently available radio diagnostics by themselves or in combination with other wavelengths provide data sets which should help us to go beyond simplistic models of large-scale coronal restructuring and its consequences.

Key questions on the instability and injection of large–scale structures in the corona where the joint application of radio diagnostics with imaging and spectrography in white light and XUV can further our understanding are:

1. What is the origin of CMEs and how do they propagate through the corona into the interplanetary space ?
2. Are flares and CMEs causally related or are they different consequences of the same physical process ? More specifically: does a CME affect the corona in such a way that energy is released whereas it would not have been released in the absence of a CME ? If so, does the CME provide the energy which is released in its aftermath (e.g. in opening the magnetic field configuration which subsequently reconnects), or does it just trigger the release of energy stored independently in active regions ?

Multi-frequency radio observations at short wavelengths give access to the state of the thermal plasma in a prominence and its surroundings. They are a *possible* means to probe the thermal emission of CMEs, providing a density diagnostic complementary to coronographic observations. Decimetric-to-metric observations localize the sites of *electron* acceleration and, by their sensitivity to even small amounts of non thermal electrons, are a tracer of energy release processes as precursors or in the aftermath of a mass ejection.

Type II radio bursts remain the principal clear signature of shock waves in the corona. In situ observations of planetary bow shocks and interplanetary shocks have furthered our understanding especially of the particle acceleration processes, but basic parameters of coronal shock waves and their environment are unknown. This makes it difficult to simply transfer this understanding to the coronal and astrophysical case:

1. Although there is plausible evidence that coronal type II emission comes from blast waves rather than from CME-driven shocks, the detailed relationship between CMEs and flares with coronal and interplanetary shocks is not fully understood. Electron acceleration and radio wave generation should not depend on the origin of the shock wave. Radio wave signatures of bow shocks of CMEs should be observable if they accelerate solar energetic particles in the high corona. The filling of the observational gap between 15 and 5 MHz by the WIND radio wave detector and the availability of coronographic observations up to 30 $R_\odot$ with SOHO open unprecedented opportunities for studying shock wave propagation over an extended height range.
2. The question where the sources of type II emission are located with respect to the shock wave is open. Available evidence is not sufficient to decide between quasi-parallel and quasi-perpendicular acceleration regions, or the upstream and downstream domain. Different substructures of the type II spectra have been suggested to come from different sites, but the evidence is at best tentative.
3. Our ignorance of the Alfvén speed in the low and middle corona remains a major obstacle to the understanding of plasma processes at coronal shock waves. Many observations can be understood in terms of blast waves being refracted into regions of high density, i.e. perhaps low Alfvén speed. It will hopefully be possible in the future to combine radio observations with vector magnetic field measurements and their extrapolation into the corona to get a better understanding.

Many of these open questions will be addressed in the coming years, since SOHO carries sophisticated instrumentation with a degree of complementarity unknown up to now. There is no doubt that these new data combined with ground based observations, e.g., radio measurements, and supported by additional theoretical investigations will lead to a better understanding of plasma processes in the solar corona.

Acknowledgements: The authors are grateful to the members of the working group *Shocks and CMEs* whose contributions are the core of the present report. They include H. Aurass, I. Chertok, N. Gopalswamy, P. Hackenberg, A. Klassen, V. Krasnoselskikh, P.K. Manoharan, P. Zlobec, E. Zlotnik. They acknowledge S. Kahler's helpful comments on the manuscript.

References

Akimov V.V. and 6 co-authors (1994): Some evidences of prolonged particle acceleration in the high-energy gamma ray flare of June 15, 1991. In *High-Energy Phenomena–A New Era of Spacecraft Measurements*, ed. by J.M. Ryan, W.T. Vestrand, AIP Conf. Proc. 294, 106

Akimov V.V. and 11 co-authors (1996): Evidence for prolonged acceleration based on a detailed analysis of the long-duration solar gamma ray flare of June 15, 1991. Solar Phys. 166, 107

Alexander D., Slater G.L., Hudson H.S., McAllister A.H., Harvey K.L. (1994): The large scale coronal eruptive event of April 14 1994. In *Solar Dynamic Phenomena and Solar Wind Consequences*, ESA SP-373, 187

Aurass H. (1992): Radio observations of coronal and interplanetary type II bursts. Ann. Geophys. 10, 359

Benz A. O., Thejappa G. (1988): Radio emission of coronal shock waves. A&A 202, 267

Bougeret J.-L. (1985): Observations of shock formation and evolution in the solar atmosphere. In *Collisionless Shocks in the Heliosphere: Reviews of Current Research*, ed. by B. S. Tsurutani, R. G. Stone (Geophys. Monogr. Ser. Vol. 35, Washington DC), 13

Bougeret, J.-L. and 11 co-authors (1995): Waves: The radio and plasma wave investigation on the Wind spacecraft. In *The Global Geospace Mission*, ed. by C. T. Russell (Kluwer Academic Press, Dordrecht), 231

Burkepile J.T., St. Cyr O.C. (1993): A revised and expanded catalogue of mass ejections observed by the Solar Maximum Mission Coronagraph. NCAR Technical Note NCAR/TN-369+STR

Cairns I. H. (1986): New waves at multiples of the plasma frequency upstream of the Earth's bow shock. JGR 91, 2975

Cairns I. H. (1987a): Fundamental plasma emission involving ion sound waves. J. Plasma Phys. 38, 169

Cairns I. H. (1987b): Second harmonic plasma emission involving ion sound waves. J. Plasma Phys. 38, 179

Cairns I. H. (1987c): Third and higher harmonic plasma emission due to Raman scattering. J. Plasma Phys. 38, 199

Cairns I. H., Robinson R. D. (1987): Herringbone bursts associated with type II solar radio emission. Solar Phys. 111, 365

Chertok I.M. (1996): Yohkoh data on CME-flare relationships and post-eruption magnetic reconnection in the corona. In *Yohkoh Conference on Magnetic Reconnection in the Solar Atmosphere*, ed. by R.D. Bentley, J.T. Mariska, ASP Conf. Series, in press

Démoulin P., Vial J.C. (1992): Structural characteristics of eruptive prominences. Solar Phys. 141, 289

Dryer M. (1994): Interplanetary studies: propagation of disturbances between the sun and the magnetosphere. Space Sci. Rev. 67, 363

Dulk G. A., McLean D. J. (1978): Coronal magnetic fields. Solar Phys. 57, 279

Fainberg J., Stone R. G. (1971): Type III solar radio burst storms observed at low frequencies. Solar Phys. 17, 392

Feynman J., Hundhausen A.J. (1994): Coronal mass ejections and major solar flares: the great active Center of March 1989. JGR 99, 8451

Filbert P. C., Kellog P. J. (1979): Electrostatic noise at the plasma frequency beyond the Earth's bow shock. JGR 84, 1369

Gopalswamy N. and 6 co-authors (1996): Radio and X-ray studies of a CME associated with a very slow prominence eruption. ApJ, submitted

Gopalswamy N., Kundu M.R. (1989): A slowly moving plasmoid associated with a filament eruption. Solar Phys. 122, 91

Gopalswamy N., Kundu M. R. (1992): Are coronal shocks piston driven ? In *Particle Acceleration in Cosmic Plasmas*, ed. by G. P. Zank, T. K. Gaisser (American Institute of Physics, New York), 257

Gosling J. T., Birn J., Hesse M. (1995): Three-dimensional magnetic reconnection and the magnetic topology of coronal mass ejection events. Geophys. Res. L. 22, 869

Gurnett D. A., Neubauer F. M., Schwenn R. (1979): Plasma wave turbulence associated with an interplanetary shock. JGR 84, 541

Hanaoka Y. and 19 co-authors (1994): Simultaneous observation of a prominence eruption followed by a coronal arcade formation in radio, soft X-rays, and H_α. Publ. Astron. Soc. Japan 46, 205

Harrison R.A., Hildner E., Hundhausen A.J., Sime D.G., Simnett G.M. (1990): The launch of solar coronal mass ejections: results from the coronal mass ejection onset program. JGR 95, 917

Hiei E., Hundhausen A.J., Sime D.G. (1993): Reformation of a coronal helmet streamer by magnetic reconnection after a coronal mass ejection. Geophys. Res. L. 20, 2785

Hildner E. and 20 co-authors (1986): Coronal mass ejections and coronal structures. In *Energetic Phenomena on the Sun*, ed. by M.R. Kundu, B.E. Woodgate, NASA CP-2439, 6-1

Holman G. D., Pesses M. E. (1983): Solar type II radio emission and shock drift acceleration of electrons. ApJ 267, 837

Howard R.A., Sheeley N.R., Koomen M.J., Michels D.J. (1985): Coronal mass ejections: 1979-1981. JGR 90, 8173

Hudson H.S. and 7 co-authors (1994): Non-thermal effects in slow solar flares. In *X-Ray Solar Physics from Yohkoh*, ed. by Y. Uchida, T. Watanabe, K. Shibata, H.S. Hudson (Univ. Acad. Press, Tokyo), 143

Hudson H.S., Haisch B., Strong K.T. (1995): Comment on "The solar flare myth" by J.T. Gosling. JGR 100, 3473

Hughes W.J., Sibeck D.G. (1987): On the 3-dimensional structure of plasmoids. Geophys. Res. L. 14, 636

Hundhausen A.J. (1987): The origin and propagation of coronal mass ejections. In *Solar Wind 6*, ed. by V.J. Pizzo, T.E. Holzer, D.G. Sime, NCAR Tech. Note/TN-306+Proc, 181

Hundhausen A.J. (1993): Sizes and locations of coronal mass ejections: SMM Observations from 1980 and 1984-1989. JGR 98, 13177

Hundhausen A.J. (1995): Coronal mass ejections: a summary of SMM observations from 1980 and 1984-1989. To be published in *The Many Faces of the Sun*

Kahler S.W., Hundhausen A.J. (1992): The magnetic topology of solar coronal structures following Mass Ejections. JGR 97, 1619

Kahler S.W., Moore R.L., Kane S.R., Zirin H. (1988): Filament eruptions and the impulsive phase of solar flares. ApJ 328, 824

Kanbach G. and 18 co-authors (1993): Detection of a long-duration solar gamma-ray flare on June 11, 1991 with EGRET on Compton-GRO. A&AS 97, 349

Kennel C. F., Edmiston J. P., Hada T. (1985): A quarter century of collisionless shock research. In *Collisionless Shocks in the Heliosphere: Reviews of Current*

Research, ed. by B. T. Tsurutani, R. G. Stone (Geophys. Monogr. Ser. Vol. 35, Washington DC), 1

Kennel C. F., Scarf F. L., Coroniti F. V., Smith E.J., Gurnett D.A. (1982): Nonlocal plasma turbulence associated with interplanetary shocks. JGR 87, 17

Klein K.-L., Mouradian Z. (1990): A Moving Radio Source Related to a Prominence Eruption. In *Flares 22 Workshop "Dynamics of Solar Flares"*, ed. by B. Schmieder, E.R. Priest (Paris Observatory Publication), 185

Kopp R.A., Pneuman G.W. (1976): Magnetic Reconnection in the Corona and the Loop Prominence Phenomenon. Solar Phys. 50, 85

Krauss-Varban D. (1989): Fast Fermi and gradient drift acceleration of electrons at nearly perpendicular collisionless shocks. JGR 94, 15,367

Kurokawa H., Hanaoka Y., Shibata K., Uchida Y. (1987): Rotating Eruption of an Untwisting Filament Triggered by the 3B Flare of 25 April, 1984. Solar Phys. 108, 251

Leikov N.G. and 16 co-authors (1993): Spectral characteristics of high-energy gamma ray solar flares. A&AS 97, 345

Lengyel-Frey D. (1992): Location of the radio emitting regions of interplanetary shocks. JGR 97, 1609

Lengyel-Frey, D., Stone, R. G. (1989): Characteristics of interplanetary type II radio emission and the relationship to shock and plasma properties. JGR 94, 159

Leroy M. M., Mangeney A. (1984): A theory of energization of solar wind electrons by the Earth's bow shock. Ann. Geophys. 2, 449

Lin R. P. and 19 co-authors (1995): A Three-Dimensional Plasma and Energetic Particle Observation for the WIND spacecraft. In *The Global Geospace Mission*, ed. by C. T. Russell (Kluwer, Dordrecht), 125

MacQueen R. M. (1980): Coronal transients: a summary. Phil. Trans. R. Soc. London, Ser. A 297, 605

Mann G. (1995): Theory and observations of coronal shock waves. In *Coronal Magnetic Energy Releases*, ed. by A. O. Benz, A. Krüger (Springer, Berlin), 183

Mann G., Claßen H.-T. (1995): Electron acceleration to high energies at quasi-parallel shock waves in the solar corona. A&A 304, 576

Mann G., Claßen T., Auraß H. (1994a): Characteristics of coronal shock waves and solar type II radio bursts. A&A 295, 775

Mann G., Lühr H., Baumjohann W. (1994b): Statistical analysis of short large amplitude magnetic structures at a quasi-parallel shock. JGR 99, 13,315

Marsch E. (1990): Kinetic physics of the solar wind plasma. In *Physics of the Inner Heliosphere* Vol. 2, ed. by R. Schwenn, E. Marsch (Springer, Berlin, Heidelberg, New York), 103

McAllister A.H., Dryer M., McIntosh P., Singer H., Weiss L. (1994): A large polar crown CME and a severe geomagnetic storm: April 14–17, 1994. In *Solar Dynamic Phenomena and Solar Wind Consequences*, ESA SP-373, 315

Melrose D. B. (1985): Plasma emission mechanisms. In *Solar Radiophysics*, ed. by D. J. McLean, N. R. Labrum (Cambridge University Press, Cambridge), 177

McLean D.J. (1973): A moving radio burst on the limb of the sun observed at 80 and 160 MHz. Proc. Astron. Soc. Aust. 2, 222

Munro R. H., Sime D. G. (1985): White-light coronal transients observed from Skylab May 1973 to February 1974: a classification by apparent morphology. Solar Phys. 97, 191

Nelson G. S., Melrose D. (1985): Type II bursts. In *Solar Radiophysics*, ed. by D.J. McLean, N.R. Labrum (Cambridge University Press, Cambridge), 333

Paschmann G., Sckopke N., Asbridge J. R., Bame S. J., Gosling J. T. (1980): Energization of solar wind ions by reflection from the Earth's bow shock. JGR 85, 4689

Priest E.R. (1982): *Solar Magnetohydrodynamics* (D. Reidel, Dordrecht)

Rompolt B. (1990): Small scale structure and dynamics of prominences. Hvar Obs. Bulletin 14(1), 37

Schwartz S. J., Burgess D. (1991): Quasi-parallel shocks: a patchwork of three-dimensional structures. Geophys. Res. L. 18, 373

Schwartz S. J., Burgess D., Wilkenson W. P., Kessel R. L., Dunlop M., Lühr H. (1992): Observations of short large amplitude magnetic structures at a quasi-parallel shock. JGR 97, 4209

Schwartz S. J., Thomsen M. F., Gosling J. T. (1983): Ions upstream of the Earth's bow shock: A theoretical comparison of alternative source populations. JGR 88, 2039

Smerd S.F., Dulk G.A. (1971): 80 MHz radioheliograph evidence on moving type IV bursts and coronal magnetic fields. In *Solar Magnetic Fields*, IAU Symp. no. 57, ed. by R. Howard (D. Reidel, Dordrecht), 616

Smerd S. F., Wild J. P., Sheridan K. V. (1962): On the relative position and origin of harmonics in the spectra of solar radio bursts of spectral types II and III. Aust. J. Phys 15, 180

Sonnerup B. U. Ö. (1969): Acceleration of particles reflected at a shock front. JGR 74, 1301

Stewart R.T. and 6 co-authors (1982): Visible light observations of a dense plasmoid associated with a moving type IV solar radio burst. A&A 116, 217

Stone R. G. and 31 co-authors (1992): The unified radio and plasma wave investigation. A&AS 92, 291

Thejappa G., Wentzel D. G. MacDowall R. J., Stone R. G. (1995): Unusual wave phenomena near interplanetary shocks at high latitudes. Geophys. Res. L. 22, 3421

Treumann R. A. and 12 co-authors (1986): Electron plasma waves in the solar wind: AMPTE/IRM and UKS observations. Adv. Space Res. 6(1), 93

Tsuneta S. (1996): Evidence for reconnection in solar flares from Yohkoh SXT observations. ApJ 456, 840

Wagner W.J., MacQueen R. M. (1983): The excitation of type II radio bursts in the corona. A&A 120, 136

Webb D.F., Cliver E.W. (1995): Evidence for magnetic disconnection of mass ejections in the corona. JGR 100, 5853

Wibberenz G. K. and 6 co-authors (1989): Coronal and interplanetary transport of solar energetic protons and electrons. Solar Phys. 124, 353

Wild J. P., Sheridan K. V., Trent G. H. (1959): The transverse motions of the source of solar radio bursts, In IAU/URSI Symp. Paris Symposium on Radio Astronomy, ed. by R.N. Bracewell (Stanford Univ. Press, Stanford), 176

Wu C. S. (1984): A fast Fermi process: energetic electrons accelerated by a nearly perpendicular bow shock. JGR 89, 8857

Part IV

Radio Instrumentation

An Upgrade of Nobeyama Radioheliograph to a Dual-Frequency (17 and 34 GHz) System

Toshiaki Takano[1], Hiroshi Nakajima[1], Shinzo Enome[1]*, Kiyoto Shibasaki[1],
Masanori Nishio[1], Yoichiro Hanaoka[1], Yasuhiko Shiomi[1], Hideaki Sekiguchi[1],
Susumu Kawashima[1], Takeshi Bushimata[1], Noriyuki Shinohara[1], Chikayoshi
Torii[1], Kenichi Fujiki[1] and Yoshihisa Irimajiri[2]

[1] Nobeyama Radio Observatory, Nobeyama, Minamisaku, Nagano 384-13, Japan
[2] The Communications Research Laboratory, Koganei, Tokyo 184, Japan

Abstract. The Nobeyama Radioheliograph, originally constructed as a 17 GHz system, was upgraded into a dual-frequency system operating at 17 and 34 GHz on a time sharing basis. For each of the 84 antennas, a frequency-selective sub-reflector, which reflects 17 GHz radio waves into the Cassegrain focus while transmits 34 GHz waves into the primary focus, was installed and a 34 GHz frontend receiver system was mounted in parallel with the existing 17 GHz system. No major modification was introduced to the backend system. Neither were antennas added nor their arrangement changed. With this minimal modification, we have obtained (1) an angular resolution of $\sim 5''$ (at 34 GHz) and (2) a spectral diagnostic capability of cm- to mm-wave emissions from solar flares with temporal resolution up to 100 ms. Daily 8-hour (from $\sim$22:45 to $\sim$6:45 UT) operation at dual frequencies started late October, 1995. Final tuning of the new system, such as the calibration and development of image synthesis software tools is still under way. Flare images taken at the dual frequencies are presented and compared with that from the *Yohkoh* SXT as an example.

Résumé. A l'origine le radiohéliographe de Nobeyama a été construit pour opérer à 17 GHz. Cet instrument a été récemment développer pour fonctionner, en temps partagé, à deux fréquences, 17 et 34 GHz. Chacune des 84 antennes a été équipée d'un réflecteur secondaire sélectif qui réfléchit le signal reçu à 17 GHz vers le foyer Cassegrain alors qu'il transmet le signal 34 GHz vers le foyer primaire. Le système de réception focal à 34 GHz a été installé en parallèle avec celui dejà existant à 17 GHz. Le reste du système de réception n'a pas été profondément modifié et aucun changement n'a été apporté aux réseaux d'antennes. Ces modifications minimales permettent d'obtenir: (1) une résolution angulaire de $\sim 5''$ à 34 GHz et (2) une possibilité de diagnostic spectral des émissions centimétriques et millimétriques associées aux éruptions, avec une résolution temporelle qui peut atteindre 100 ms. Les observations journalières, d'une durée d'environ 8 heures ($\sim$22:45 – $\sim$6:45 TU), ont commencées fin-octobre 1995. Des réglages finaux du système, par exemple pour la calibration, et des développements de logiciels pour la reconstitution des images de synthèse sont en cours de réalisation. Des images d'éruptions obtenues aux deux fréquences sont montrées à titre d'exemples et comparées aux images X-mou obtenues avec l'instrument *Yohkoh* SXT.

---

* also at VSOP office, National Astron. Obs., Mitaka, Tokyo 181, Japan

1   Introduction

The Nobeyama Radioheliograph was originally constructed as a 17 GHz interferometer with 84 0.8 m antennas arranged in a redundantly-spaced T-shaped array (Nakajima et al. 1994). Each pair of antennas measures a complex Fourier component of the brightness distribution of the Sun. By inverse Fourier transformation, the radioheliograph provides a pair of right- and left-handed circular polarization images of the whole Sun once every second during routine observations and every 50 ms for selected events such as flares. The spatial resolution is moderately high, ~ 10″, as a full-Sun imaging instrument. One more advantage of the radioheliograph is its high image quality. The gain and phase of each antenna are calibrated from observations of the Sun itself ("self calibration") based upon the redundancy of the antenna baselines. This calibration method realizes rather high accuracy in measuring the Fourier components, so that snap-shot images of a large flare with a dynamic range of $\gtrsim 25$ dB can be synthesized from data taken with 1 s integration time (Koshiishi et al. 1994). Recently, Koshiishi (1996) developed a new software with application of the Steer algorithm in CLEAN procedure (Steer et al. 1984); with this software faint features of a few hundred Kelvin can be synthesized against the solar disk of 10000 K in snap-shot images with 10 s integration. With these advanced capabilities, a variety of interesting phenomena have been, currently are, and will be, investigated: see, e.g., reviews by Enome (1995, 1996) and references therein.

Single-frequency observations at 17 GHz, however, are not enough to derive physical quantities involved in the cm-wave emission sources; spectral information is crucial in this regard. We therefore decided to upgrade the radioheliograph into a dual-frequency system. A higher frequency is suitable for the secondary frequency, because we obtain correspondingly higher spatial resolution and, further, because the combination between 17 GHz and a higher frequency provides an important constraint on the spectrum of high-energy electrons that radiate optically-thin gyrosynchrotron emission in this frequency range. We chose 34 GHz, because by adopting this doubled frequency we can make use of most of the existing hardware in common for both frequencies. Note also that the terrestrial atmosphere is relatively transparent at this frequency and further that we may expect some drastic change of radio images between 17 GHz and 34 GHz, at least for some types of sources; e.g., the gyroresonance emission from thermal electrons above sunspots becomes less important at 34 GHz, while it is rather strong at 17 GHz.

This upgrade was conducted during the fall of 1995. In this paper, we briefly describe the essence of the dual-frequency system and discuss its performance in Section 2. An example of flare observations is shown in Section 3. A full description of the instrument and discussion on its total performance will be made elsewhere in the near future (Takano et al., in preparation).

Table 1. Parameters of the Nobeyama Radioheliograph

Basic Parameters		
Number of antennas	84	
Diameter of antenna	0.8 m (=45λ at 17 GHz)	
Full Baseline lengths	EW 488.960 m (=27708λ at 17 GHz)	
	NS 220.060 m (=12470λ at 17 GHz)	
Fundamental spacing (d)	1.528 m (=87λ at 17 GHz)	

	17GHz System	17/34GHz System	
		17 GHz	34 GHz
Observing Frequency	17.0 GHz	17.0 GHz	33.8 GHz
Wave Length	17.6 mm	17.6 mm	8.9 mm
Spatial resolution	~ 10″	~ 10″	~ 5″
Time resolution	50 ms	100 ms	100 ms
Polarization	RCP and LCP	RCP and LCP	Linear

2   The Dual-Frequency System

All 84 of the existing antennas are used for the dual-frequency imaging and neither new antennas were added nor the arrangement of antennas in the array was changed in this upgrade to the dual-frequency system. The key instrumental parameters are given in Table 1.

As shown in Table 1, the spatial resolution at 34 GHz is ~ 5″, twice as good as that at 17 GHz. No capability for circular polarization measurement was added at 34 GHz in this upgrade. The fundamental spacing between antennas (1.528 m) was designed to obtain the Fourier component which corresponds to 40′ at 17 GHz so that full-Sun maps free from the aliasing effect are taken in snap-shot images. The situation differs at 34 GHz : when we are interested in global, stationary structures on the Sun, the aliasing effect needs be removed by a super-synthesis technique. This aliasing effect is not a serious problem, however, for observing rapidly varying flare sources: they usually appear in an active region and are far brighter than other quiet components so that they can easily be distinguished from the background or, if necessary, the pre-flare brightness distribution can be easily subtracted.

A schematic diagram of the dual-frequency system for one antenna is shown in Figure 1. Note that the dual-frequency observations are carried out on a time-sharing basis (Fig. 2) by switching the intermediate frequency (IF) signals. The highest time resolution in the dual-frequency operation is, therefore, 100 ms, which is two times worse than that in the single-frequency

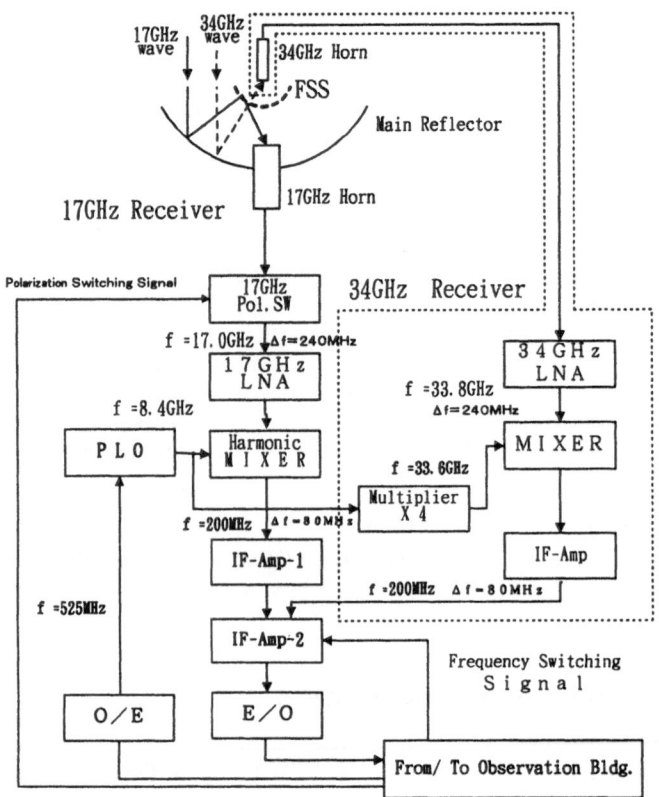

Fig. 1. Schematic diagram of the dual-frequency system (for each antenna). A frequency selective sub-reflector (FSS) and a 34 GHz receiver, enclosed by a dotted line, were newly installed.

mode of operation. A part of the IF and the backend system are used in common for the dual frequencies without any significant modifications. In the following, we will discuss essential points of the new system.

2.1   Frequency-Selective Sub-Reflectors and Antenna Optics

In order to accommodate the dual-frequency signals separately, frequency-selective sub-reflectors (FSS's) are mounted on the 84 antennas replacing the original aluminum sub-reflectors. The FSS reflects 17 GHz radio waves back to the Cassegrain focus while transmits 34 GHz waves into the primary focus, where a 34 GHz horn is newly placed.

The FSS's are made of gold-plated fine ceramics (Figure 3). The fine ceramic FSS's have quite good performance against heat cycles, sunshine,

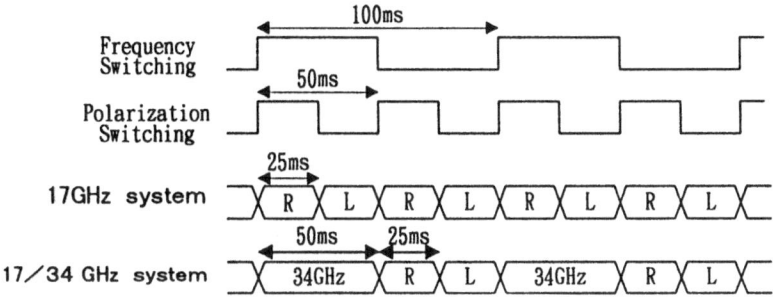

Fig. 2. Time-sharing sequence of observations.

water and ice, etc. Jerusalem cross patterns are drawn on the gold plate by chemical etching for the selection of frequencies. These patterns are etched only in the inner half of the sub-reflector surface, since the 34 GHz horn illuminates only the inner half of the main-reflector in order to obtain roughly the same primary beam size ($\sim 85'$ in FWHM) as that at 17 GHz.

2.2   34 GHz Frontend Receivers

A 34 GHz frontend receiver, consisting of a low noise HEMT amplifier, a mixer, and an IF amplifier, was installed in the frontend box on each antenna. In this receiver the 34 GHz signal is down-converted to an IF signal at 200 MHz, which is the same IF frequency for 17 GHz. The 33.6 GHz local frequency signal for this down-conversion is provided by the 8.4 GHz phase-locked oscillator (PLO), which is also used for the down-conversion of the 17 GHz signal. The down-converted signal is amplified and transmitted to a diode switch which selects the 17 or 34 GHz input signals.

2.3   Fringe Stopping and Delay Adjustment

In the observing building, the 200 MHz IF signals from all of the 84 antennas are down-converted again into the baseband by being mixed with a second local frequency signal at 200 MHz. The fringe stopping is made here by rotating phases of the second local frequency signal, where the phase rotation rates differ for one antenna to another, depending on the locations of the individual antennas in the array. The rates are calculated for the 17 GHz signals. For stopping fringes of the 34 GHz signals, phase rotation rates just twice as large as those for 17 GHz are applied. This does not exactly stop the fringes, since the ratio between the two observing frequencies (33.8 GHz *vs* 17.0 GHz) is not exactly 2. The remaining rotation of the phase makes no

Fig. 3. Frequency-selective sub-reflector.

significant degradation of images because the phases are calibrated for every 1 s data with the self calibration method during which the rotation is slow enough.

The difference of delay between the 17 and 34 GHz signals is mainly caused by separate signal paths in the frontend receivers. This difference was measured to be constant for all antennas within 1 ns. This amount of delay, 1 ns, causes only 0.2% loss of correlation amplitudes for 34 GHz signals if the delay for 17 GHz signals are well adjusted. We, therefore, use the common values of the delay for both frequency signals of individual antennas to compensate the arrival times of signals at the correlator.

2.4   Basic Performance

Some parameters that are relevant to the performance of the upgraded, dual-frequency system are summarized in Table 2. Note that no degradation in the dual-frequency system was found in the 17 GHz part except for the sensitivity reduction due to the time sharing. The 34 GHz part has as good a performance as the 17 GHz part except for the receiver noise temperature. Examination of the overall performance of the dual-frequency system as an imaging instrument is now under progress.

Table 2. Basic Performance of the System (Typical Values)

	17GHz System (old system)	17/34GHz System 17 GHz	34 GHz
Receiver Noise Temperature	~360 K	~360 K	~1000 K
Sensitivity in Images of 1-sec Full-UV Snapshot[1]	4.4×10^{-3} sfu $(700\,\text{K})^2$	6.2×10^{-3} sfu $(1000\,\text{K})^2$	12.2×10^{-3} sfu $(6000\,\text{K})^2$
Isolation between 17, 34 GHz	—	$\lesssim -50\,\text{dB}$ ($34{\rightarrow}17\,\text{GHz}$)	$\lesssim -50\,\text{dB}$ ($17{\rightarrow}34\,\text{GHz}$)
Crosstalk Level between Antennas	$< -50\,\text{dB}$	$< -50\,\text{dB}$	$< -50\,\text{dB}$
Degree of Cross-Polarization	$\lesssim -20\,\text{dB}$	$\lesssim -20\,\text{dB}$	—

[1] Calculated with the measured receiver noise temperature.
[2] For compact sources.

3   Example of Flare Observations

Images of a flare observed on November 10, 1995 are shown in Figure 4. This flare, which started at 03:41UT in NOAA 7921 and lasted for about 100 min, is an LDE flare. Radio images were synthesized with newly-developed software by Fujiki (1996) for higher spatial-resolution imaging, in which only UV data at multiples of 8 or 16 times the fundamental spacing (d) are used instead of the full UV data. This software produces partial-Sun images with the field of view corresponding to 8d or 16d spacing, and can avoid superposition of sources outside the field of the image.

Differences in the spectrum as well as in the circular polarization of the two radio sources is remarkable: the northern source is unpolarized at 17 GHz and is seen in all of the four images in Figure 4, while the southern source is seen clearly only in the left-handed circular polarization (LCP) image at 17 GHz (degree of polarization $\sim$ 85%). The brightness temperature of the northern source is 2.8×10^5 K at 17 GHz (both in RCP and LCP) and 8.7×10^4 K at 34 GHz. The flux of the northern source is about 3 sfu at both frequencies. The southern source has the brightness temperature of 9.9×10^5 K at 17GHz in LCP. This source is located on the umbra of the largest sunspot.

190   Toshiaki Takano et al.

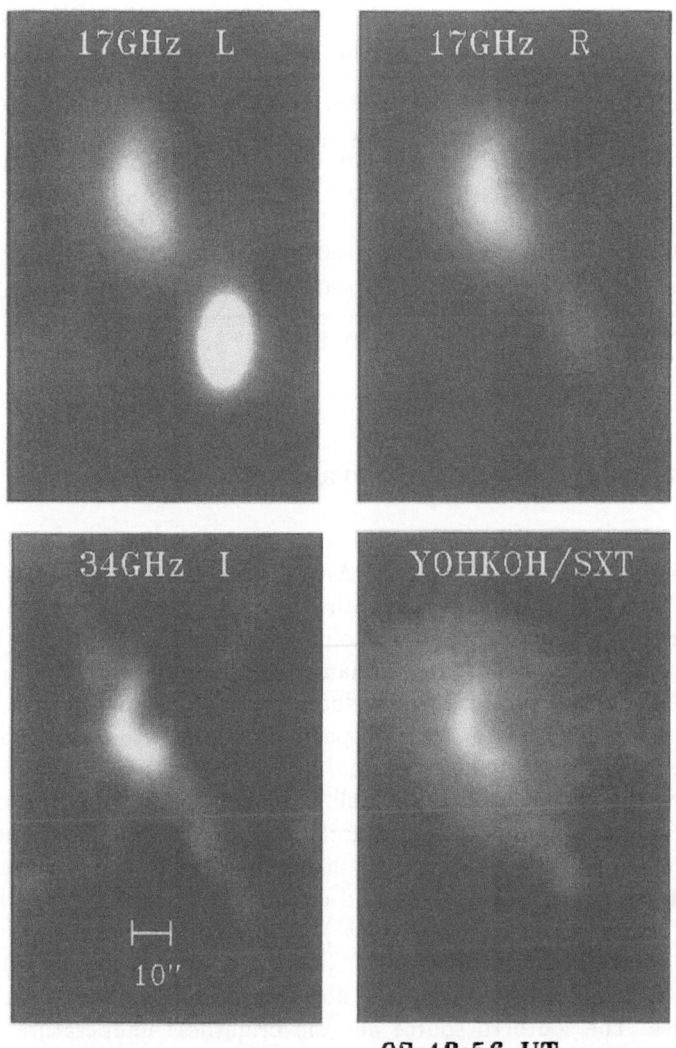

03:48:56 UT

Fig. 4. Images of a flare of November 10, 1995. Left- and right-handed circular polarization images at 17 GHz and a 34 GHz image are compared with a soft X-ray image from *Yohkoh* SXT. Integration time for radio images is 60 s.

Because the bright part of the northern source shows a similar shape to the soft X-ray image (angular resolution is 2.5″) and the 34 GHz image, the source is resolved with the spatial resolution of 5″ beam at 34 GHz, but is unresolved by the 10″ beam at 17 GHz. The northern source may be interpreted as free–free emission from a thermal plasma filling a magnetic loop, since it is unpolarized, shows a flat spectrum, and resembles the soft X-ray bright loop-like structure. On the other hand, the southern source may be interpreted as gyroresonance emission in strong magnetic fields above the sunspot, since it has quite a steep spectrum and is strongly polarized.

The observations of the November 10, 1995 flare clearly demonstrate that dual-frequency observations with the Nobeyama Radioheliograph is quite useful in clarifying the mechanisms of radio emission. We believe that the continuous operation of this instrument as the coming solar maximum approaches will provide crucial information on solar flare radio emission mechanisms and further on acceleration, confinement, and dissipation of energetic electrons.

The authors would like to thank N. Futagawa, S. Akasaka, T. Takabayashi, H. Shinohara, A. Moroi, and N. Imai of NEC Co. Ltd, S. Fujita, H. Morikawa, and A. Mori of Akasaka Diesel Co. Ltd, K. Miyatake, M. Miura, T. Yogo, E. Kanaya, T. Hirano, N. Yamada, and M. Hattori of Noritake Co. Ltd, and T. Matsuoka of Kenseido Co. Ltd. for their great contributions to realize the upgraded system. T. Kosugi gave useful comments to accomplish the manuscript.

References

Enome S., (1995): Initial Results from the Nobeyama Radioheliograph. Proc. CESRA Workshop on Coronal Magnetic Energy Releases, eds. A.O. Benz and A. Krüger, *Lecture Notes in Physics* **444**, 35–53

Enome S. (1996): A Summary of Three-Year Observations with the Nobeyama Radioheliograph. Proc. The High Energy Solar Physics Workshop, eds. R. Ramaty, N. Mandzhavidze, and X.-M. Hua, *AIP Conference Proc.* **374**, 424–432

Fujiki K. (1996): Doctor thesis, School of Mathematical and Physical Science, the Graduate University for Advanced Studies.

Koshiishi H., Enome S., Nakajima H., et al. (1994): Evaluation of the Imaging Performance of the Nobeyama Radioheliograph. *PASJ* **46**, L33–L36

Koshiishi H. (1996): Doctor thesis, School of Science, University of Tokyo.

Nakajima H., Nishio M., Enome S., et al. (1994): The Nobeyama Radioheliograph. *Proc. IEEE* **82**, No.5, 705–713

Steer D.G., Dewdney P.E., and Ito M.R. (1984): Enhancements to the Deconvolution Algorithm "CLEAN" *A & A* **137**, 159–165

The Nançay Radioheliograph

Alain Kerdraon[1] and Jean-Marc Delouis[1]

DASOP CNRS URA 2080 Observatoire de Paris, 5, place Janssen,92190 Meudon France

Abstract. The Nançay Radioheliograph has been upgraded in order to provide high time resolution 2D images of the soler corona. This is due to its new digital correlator (Stokes I and V, 576 channels). Calibration has also been improved. The multifrequency operation remains almost unchanged. It allows simultaneous observations of up to 10 frequencies in the range 150-450 MHz. Integrated data will be available on the internet, and special observing modes may be decided on request.

Résumé.
    Le Radiohéliographe de Nançay a maintenant la possibilité de faire à grande vitesse des images à 2 dimensions de la couronne solaire. Un nouveau corrélateur digital (576 canaux, Stokes I et V) a été construit à cet effet. Le mode multi-fréquence n'a pratiquement pas été modifié. Il permet l'observation quasi simultannée de 10 fréquences dans la bande 150-450 MHz. Des observations intégrées seront disponibles sur internet, et des modes d'observation speciaux pourront être mis en place à la demande

1   Introduction

The Nançay Radioheliograph (NRH) has been recently upgraded by switching from an analog to a digital correlator. The new digital correlator has 576 channels, and allows fast 2D imaging. Up to now the NRH provided only fast 1D projections of the solar corona along two axis (?). An aperture synthesis mode was also operating (?). The old analog correlator provided only 55 channels, and was at the origin of the 1D imaging limitation. It will be described in more details hereafter. Daily observations began in July 1996.In the following we briefly describe the antennas, the broadband receiver, the new correlator and the data acquisition system. Some examples of new observations will be shown.

2   Technical Description

The NRH was designed for fast imaging of solar radio emissions. Almost simultaneous observations at different frequencies are achieved by a fast time sharing (?). According to scientific and technical constraints, the observing frequencies and the speed may be chosen in the limits 150 - 450 MHz and

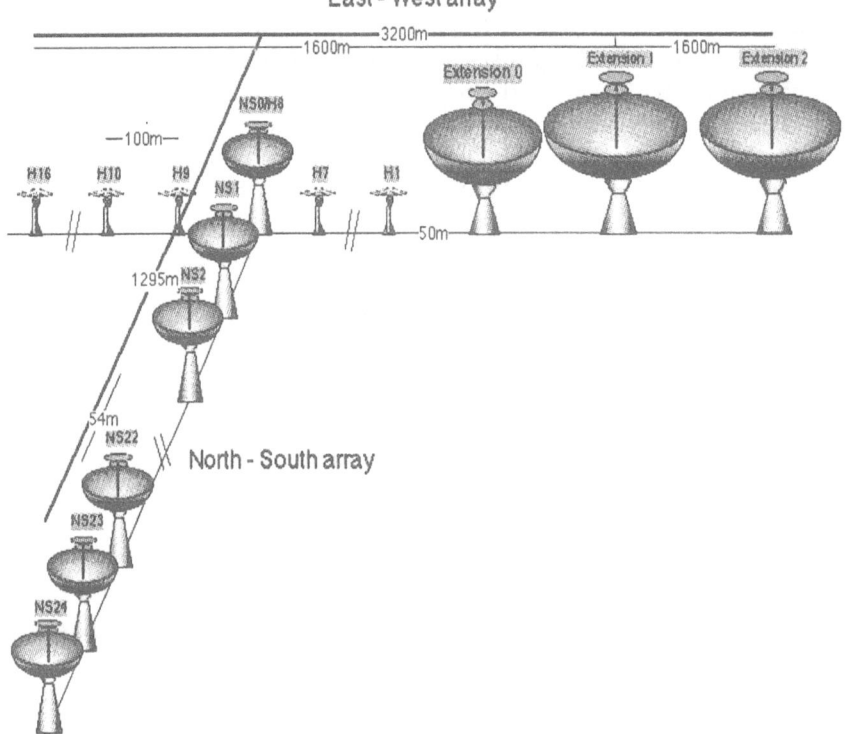

Fig. 1. The Nancay Radioheliograph antennas

a maximum number of 200 images per second. Four basic parts will be described: the antennas, the broadband receiver, the correlator, and the data acquisition system.

2.1   The Antennas

The antennas are organized in two perpendicular arrays (Fig 1). The east west array consists of 19 antennas, and provides baselines ranging from 50m to 3200m. Four antennas have parabolic collectors (two 10m dishes, one 7m dish and one 5m dish at the array crossing point). These antennas are equipped with the wide band feeds which have been described in ?. These feeds consist in four thick dipoles, and receive two orthogonal linear polarization in the 150 - 450 MHz frequency band. The remaining 15 antennas are based on a variation of thick dipoles and have no collector. They have only one linear polarization, and their gain ranges from 10 to 14 dB.

The north-south array consists of 5m parabolic collectors, with the wide band feeds described in ?. Baselines range from 54.3 to 1248m.

In order to measure circular polarization, correlations are made between signals coming from antennas with parallel dipoles(Stokes I correlations) and with perpendicular dipoles (Stokes V correlations). These measurements provide the parameter V only if the received emissions have no linear polarization. This is a very reasonnable assumption, because the Faraday rotation in the solar corona is high enough to wash away any linear polarization in the 700 kHz receiving bandwith. Very low crosstalk between perpendicular dipoles located on a same antenna has been achieved(?) and the tracking accuracy of the antennas is 1 °. This allows the measurement of polarization rates as low as a few percent.

In order to make fast 2D images with a Fourier transform, only the 17 antennas of the western part of the east - west array are used. They form, together with the north - south antennas, a T shaped array which gives a regular coverage of the u-v plane. A two dimensional Fourier transform therefore gives images which have a spatial resolution fourfold lower in the east - west direction than in 1D east -west scans. Another consequence of the arrays configuration is that the minimum east - west baseline used for 2D images is 100m. That length leads to an image size of the order of 30 arcmin at the highest frequencies, and to a strong aliasing on images. In most cases, these limitations can be by-passed by looking at 1D east - west scans.

2.2   The Broadband Receiver

Each antenna has broadband amplifiers and filters. Because of strong terrestrial interferences, a filter removes the band 200 - 210 MHz, and pass band filters reject the FM band (88 - 108 MHz) and TV emitters located very near the top edge of the receiver band (above 450 MHz). This set of filters is necessary in order to avoid intermodulation lines and saturation. A first mixer changes the observing frequency to a fixed 113 MHz frequency, which is transmitted to the laboratory through buried coaxial cables. The variable local oscillator is transmitted to the antennas also through buried coaxial cables. It is multiplied by 80 before entering the mixer, and the upper or lower image bands are used in order to reduce the frequency range of the local oscillator. The receiver of the north-south antennas is located in the antenna pedestals. Temperature variations induce a gain variation which will be corrected during the data processing (typical variations are of the order of 10% during an observing day). The receivers of the east - west antennas are buried, and do not have this drawback. For both north - south and east - west receivers, the phase is stable. Phase variations occur only in the transmission coaxial cables (signal and local oscillator). Because these cables are buried at a depth of 1m, the temperature variations are very slow, and phase calibrations remain valid for weeks.

The multifrequency operation is achieved by time sharing: a basic 5 ms integration time is used, and the observing frequency may be changed between

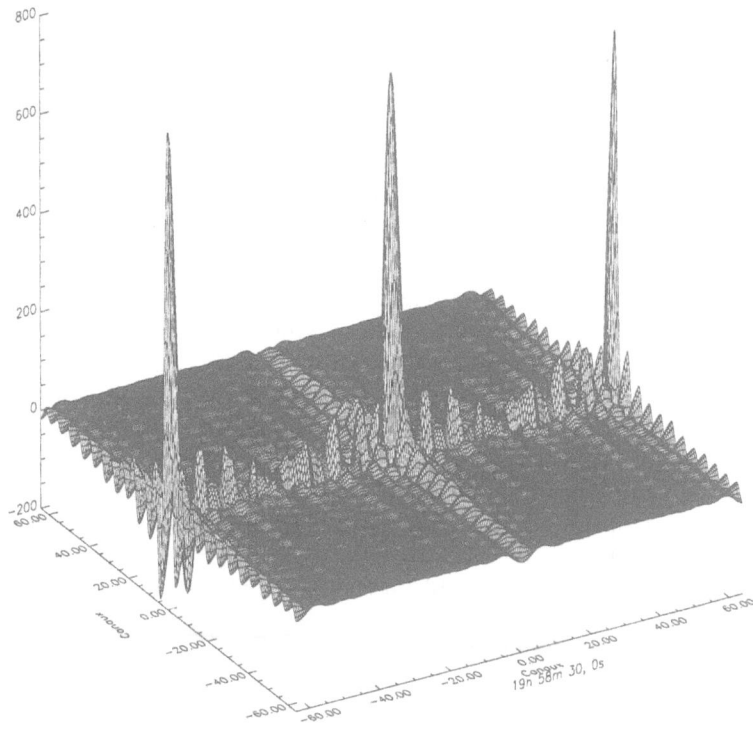

Fig. 2. Cynus A image at 164 MHz. The two horizontal axis are instrumental spatial coordinates. The vertical axis is brightness, in arbitrary units. Two periods of the image are plotted along the east-west direction.

two integrations. That change, and the needed dipoles and filters commutations are achieved during a 100 microsecond reset time. The electronic gives no limit to the number of frequencies which can be successively observed. Frequencies are chosen with a 100 kHz step, which is much smaller than the 700 kHz observing bandwith. That feature is useful because terrestrial interferences are mostly line emissions.

2.3   The Correlator

The old analog correlator provided only 55 correlations and, because of that, it was not used for fast 2D imaging. It has been replaced by a new digital correlator which is able to make all the possible correlations between the antennas. The total number of correlations is 861. Because this total includes a large number of redundant correlations, mainly in the directions of the two arrays, the correlator has been limited to 576 correlations. It makes all the products of north - south by east - west antennas, and some redundant products, mainly on short baselines in the direction of the two arrays. We used programmable logic arrays, which leads to a very compact receiver.

Before entering the digital correlator, the 113 MHz signals coming from the antennas are shifted to 10.7 MHz by a second local oscillator. Two main analog components remain unchanged: a 700 kHz passband filter determines the observing bandwidth, and an attenuation subsystem keeps the power level within a 3dB range. Attenuators range from 0 to 45 dB, with a 3dB step. They are controlled by an ancillary total power receiver. During each 5 ms period, that receiver observes at the frequency which will be used by the correlator during the next 5 ms cycle. It uses an antenna which has two independent wide band receivers, and requires the transmission of a second local oscillator to that antenna. This subsystem is needed because the solar emissions present variations, which are too strong for both the analog 10.7 MHz receiver and the analog to digital converters. Because the signals are still varying with time, a 1 bit digitization has been chosen. A separate measure of the power in the intermediate frequency is therefore needed in order to compute visibilities from the correlations. It is made by detectors whose output is integrated during 5 ms and converted to 16 bit numbers. The intermediate frequency is converted to 1 bit numbers before entering the correlator. The sampling rate has been chosen very high given the Nyquist frequency corresponding to the 700 kHz bandwidth. The sampling frequency is 42.8 MHz. That allows a very simple method to make the 90°phase shift used for sine correlation (a 1 sample shift), and the delay line time resolution (4 samples). The sampling rate is decreased before the correlation by dropping unused samples. It is divided by 13, because that leads to an over sampling close to 2 which gives, in the case of 1 bit correlations, a significant increase of the signal to noise ratio.

Table 1. Main caracteristics and observing mode

Time resolution	5ms
Spatial resolution	from 0.3 to 6 arcmin, depending on the frequency and direction
Observing frequencies	up to 10 (150 - 450 MHz)
Polarization	Stokes I and V
Observing time	7.5 h centered around 12 UT

A basic correlator cell is made of two complex correlators. As mentioned above, it computes for each pair of antennas, the correlations of parallel and perpendicular dipoles. In order to transmit the signals from two dipoles in

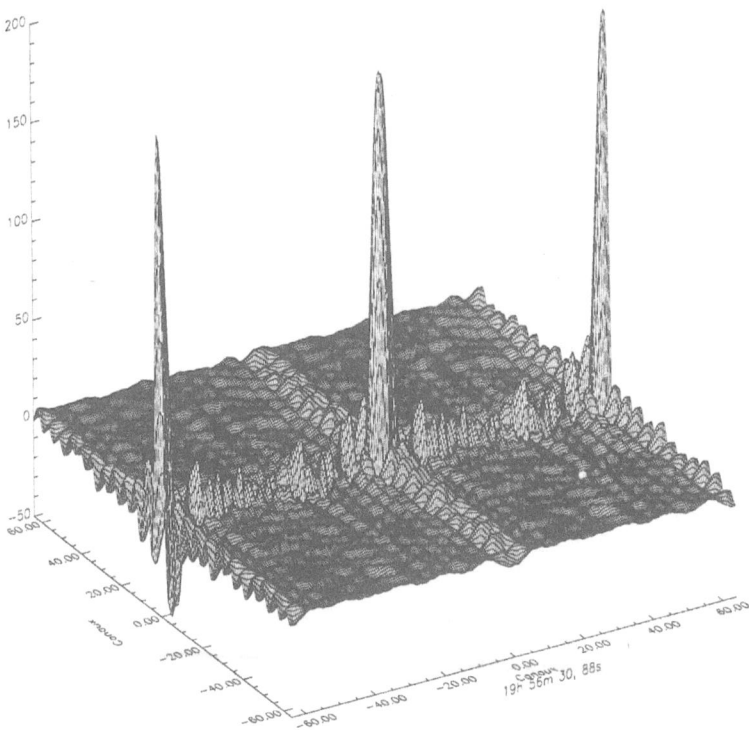

Fig. 3. Cynus A image at 432 MHz. The two horizontal axis are instrumental spatial coordinates. The vertical axis is brightness, in arbitrary units. Two periods of the image are plotted along the east-west direction.

one coaxial cable and IF electronic, a phase modulation by a set of Walsh functions is used. The phase modulation is made at the antennas, and the demodulation in the correlator. Because the NRH has many antennas and the integration time is small, the frequency of these functions is higher than usual, and an accurate knowledge of the propagation time of modulation commands and of the modulated signals is needed. This modulation scheme also reduces the effects of crosstalk on the correlations.

2.4   The Data Acquisition System and Calibration

The receiver outputs (correlations and power measurement) are fed into a real time computer (a Concurrent Computer Maxion biprocessor). This computer has two main functions: the receiver control - mainly the multifrequency operation- and the data acquisition, including real time computations. These computations include fringe stopping, computation and calibration of the

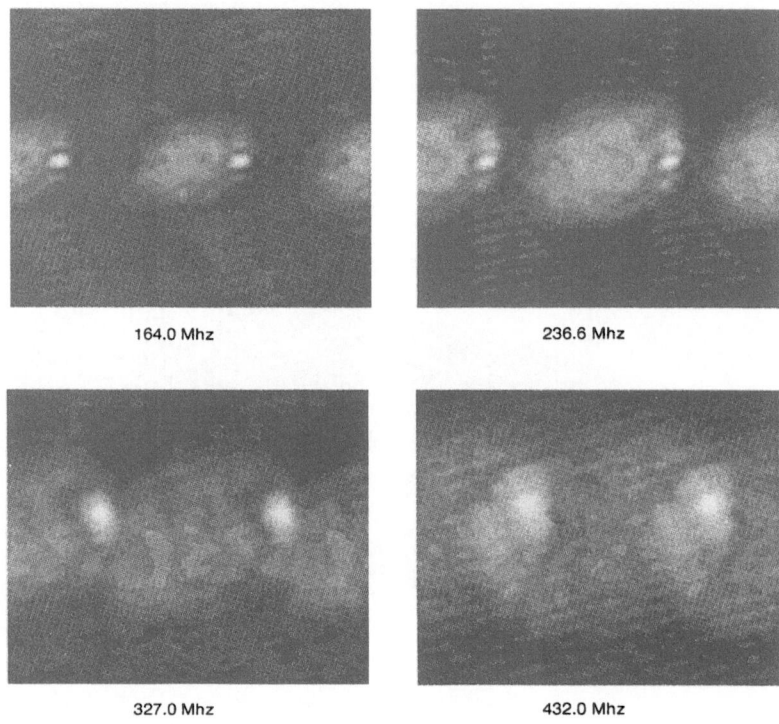

Fig. 4. The sun on 1996,July 12 at 13 32 44 UT. From top left to bottom right, four images at four frequencies are plotted (164, 236.6, 327 and 410.5 MHz). Images are computed on two periods, in order to show the aliasing phenomena at the highest frequencies.

visibilities, integration and data recording on disk. 1D and 2D raw images are also computed and displayed in real time at a low speed in order to control the proper operation of the whole system, and to allow real time scientific interaction.

The system is calibrated by observations of known intense radio sources, usually the radiogalaxy Cygnus A. Since the most intense sources are extended, a source model is fitted to the observations. Gain and phase errors are computed for each antenna and are used to compute visibilities in real time. In order to calibrate the polarization measurement, relative gain and phase of perpendicular dipoles are measured for the antennas equipped with such feeds. Because there are no polarized calibrators, the antenna at the crossing point of the arrays and two small antennas of the east - west array are rotated by 45°. Unpolarized sources therefore produce non-zero Stokes V correlations, which allow the determination of phase errors and gain. These

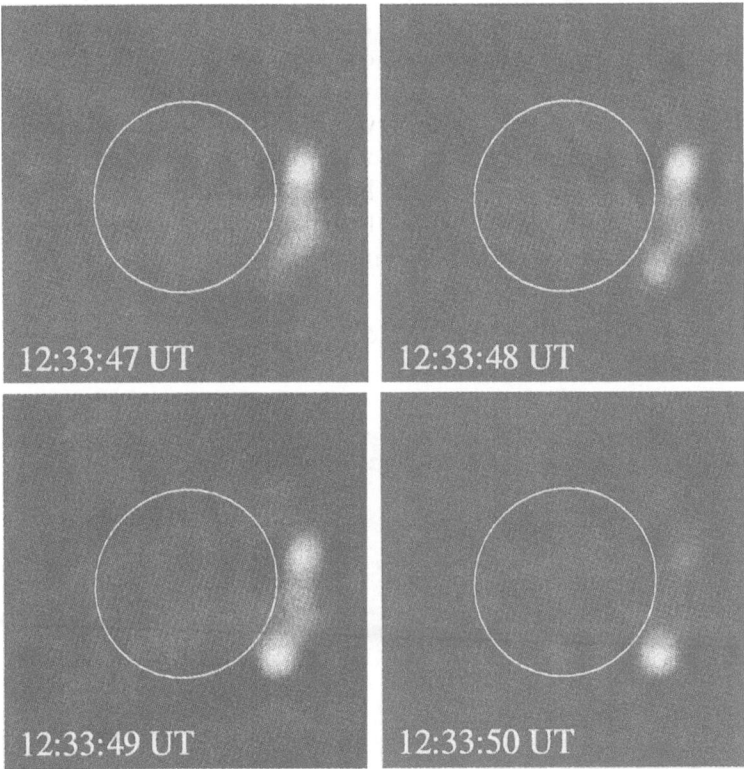

Fig. 5. The sun on 1996,July 1 at 13 33 47 UT. From top left to bottom right, four successives times are plotted, showing fast spatial evolution of type III/type V radio burst.

parameters are likely to be stable because they are determined by few electronics components and small coaxial cables located close to the dipoles. The main cause of phase error -the long buried coaxial cables- are common to both dipoles.

Figures 2 and 3 show 2D observations of Cygnus A at two frequencies. The images are very close to a theoretical image of a point source. 1D images have almost the same quality.

A standard observing mode has been defined. The major constraint is the amount of data. A limited set of visibilities (east - west and north south) are stored at 10 sets per second. Each set comprises 5 observing frequencies and Stokes parameters I and V. This set is used to compute 1D images. The total set of visibilities is stored twice per second. It is used to compute 2D images. The total amount of data is of the order of 2 Gbyte per day, for an observing time of 7,5 hours.

The frequencies are chosen in order to cover almost all the 150-450 MHz spectrum. Moreover, frequencies 164 and 327 MHz are almost mandatory because they are used for systematic observations leading to a list of radio emissions in Solar Geophysical data, and to images stored in the BASS2000 solar data base located in Paris Observatory (http://mesola.obspm.fr)

Furthermore, special observing modes will be possible on request. The main limits of such modes will be the huge amount of data generated at the maximum speed (more than 3 Gbyte/hour) and the interferences which can severely limit the choice of the observing frequencies.

3   Some Typical Observations

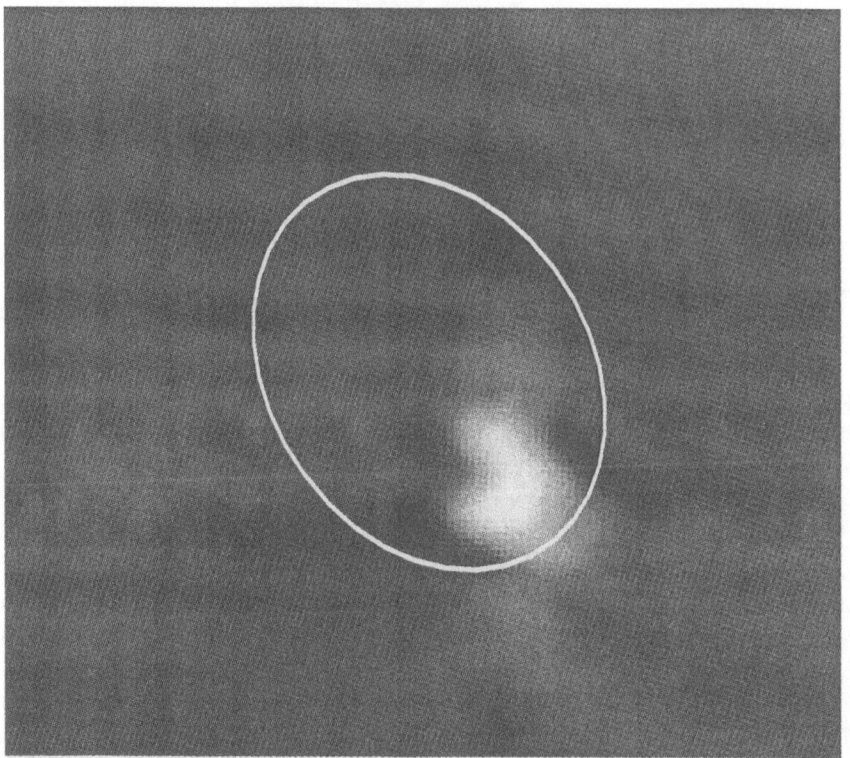

Fig. 6. The sun on 1996,July 09 at 092054 UT at 164 MHz. The white ellipsis is the optical sun limb, plotted with the geometrical distorsions which affect the raw radio map.

We will present here some observations which clearly specify the new capabilities of the NRH. Figure 4 shows an observation of a weak noise storm activity superimposed on the quiet sun. Such an emission is difficult to see on 1D projections, because the brightness contrast between the noise storm and the quiet sun is very low on such images. The non thermal source appears in figure 4 at the four plotted frequencies. These are raw images without any geometrical corrections. Two periods of the FFT image are plotted, in order to show clearly the aliasing which occurs at the highest frequencies It should be noted that the ambiguities between sources at east or west limbs can be removed by a look at 1D images, which are not affected by aliasing.

Figure 5 shows a type III/type V radio burst on 1996, July 1. Although the four successive images are displayed whith a time resolution of only 1 sec., they show that the emission consist in succesive brightenings at different positions in a loop structure. This emission is associated with a CME event.

Figure 6 shows a loop shaped continuum emission which occurred on 1996 July 9. The 2 dimensional shape of the radio source is clearly visible, although it is observed close to the lowest available frequency, which leads to the lowest spatial resolution.

4   Conclusion

The Nançay Radioheliograph has been greatly improved by its new digital correlator. The main new capability is fast 2 dimensional imaging: it is now possible to observe the 2 dimensionnal shape of radio sources which present very fast time variations. Almost simultaneous observations are available at up to ten frequencies in the 150-450 MHz frequency range. This range corresponds roughly to a 0.1- 0.5 solar radius altitude range above the photosphere. The instrument is also able to measure the circular polarization rate.

Systematic daily observations with the new receiver began on July 1996. They show that the 2 dimensionnal structure of radio bursts is well observed. A standard observing programm is used for these daily observations. It includes 1D and 2D images at 5 frequencies, respectively at 10 and 2 images per second. Integrated daily observations will be available on line, and non standard observing modes can be studied on request.

References

Alissandrakis C.E., Lantos P., 1985, Solar Physics 97, 267
The Radioheliograph Group, 1989, Solar Physics 120, 193
The Radioheliograph Group, 1993, Adv. Space Res. 13(9), 411

Recent Developments of the Solar Submm-Wave Telescope (SST)

Pierre Kaufmann[1], Joaquim E.R. Costa[2], Emilia Correia[2], Andreas Magun[3], Kaspar Arzner[3], Niklaus Kämpfer[3], Marta Rovira[4] and Hugo Levato[5]

[1] UNICAMP-NUCATE-CRAAE, Brazil
[2] INPE-CRAAE, Brazil
[3] University of Bern, IAP, Switzerland
[4] IAFE, Argentina
[5] CASLEO, Argentina

Abstract. With the approval of funds for the development and construction of the new Solar Submm-wave Telescope, its design was reviewed, updated and manufacturers for the subsystems were selected. The principal characteristics of the system and the proposed research programs are described in this note.

Résumé. Après l'approbation des resources financières pour le développement et la construction du nouveau Telescope Solaire pour ondes Sub-millimétriques, le project a eté revu, actualisé. Les constructeurs des sous-systémes ont eté selectioné. Dans cette note nous presentons les caracteristiques principales du système et les programmes de recherches proposés.

We present the basic technical characteristics of the Solar Submm-wave Telescope (SST Project). It is the first instrument designed to observe the solar continuum radiation at submm wavelengths on a regular basis. The two operating frequencies have been selected in the center of atmospheric windows at 210 GHz and 405 GHz. Special attention will be paid to the observation of solar flare emission, with high sensitivity (about 0.1 s.f.u.) and high temporal (1 ms) resolution. A compromise is made between these specifications and the antenna diameter of 1.5 m which produces beams covering a typical solar active region of a few arcminutes size. At 210 GHz a multiple receiver focal array will produce partial overlapping beams, allowing the determination of positions of flare emission centroids with an angular accuracy of a few arcseconds. This principle has already been used successfully for solar observations at 48 GHz (Georges et al., 1989; Herrmann et al.,1993; Costa et al., 1995). At 405 GHz two receivers will be used, in beamswitching mode observations. One of these receivers will have an additional intermediate frequency channel for the simultaneous measurement of an atmospheric ClO line at 390 GHz in absorption against the solar emission.

Figure 1 summarizes the principal components of the SST. It will be placed on the roof of the control room, with the antenna positioner supported by a concrete pillar rooted in the rock underlying the building. For weather and heat radiation protection the SST will operate inside a thermally

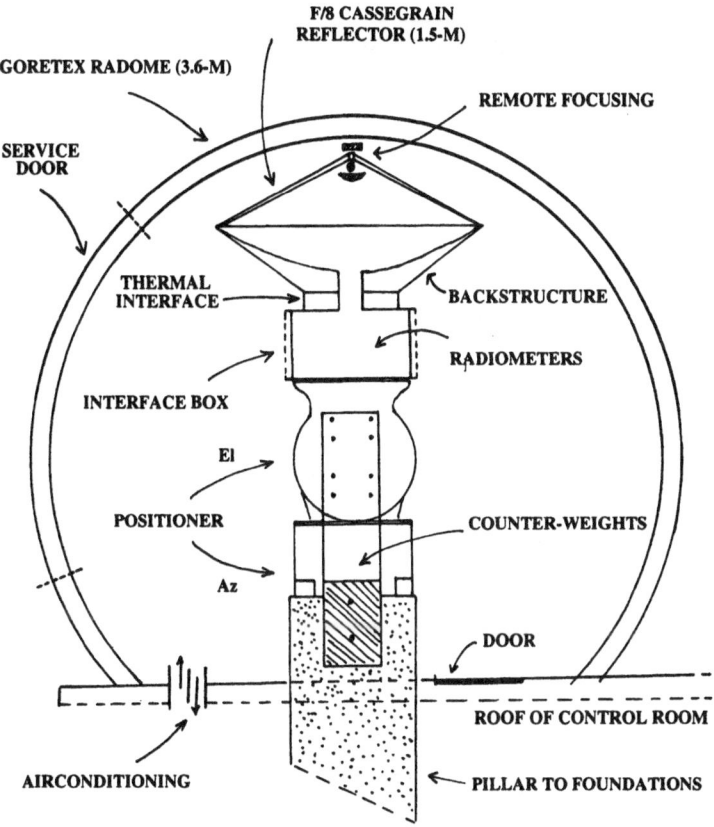

Fig. 1. Schematic diagram showing the main SST subsystems.

controlled radome. Figure 2 shows the geometrical setup of the receivers, as seen from the top. A motor-controlled mirror is used to select the signal from the antenna or the two calibration loads. The two observing frequencies are separated from the incoming signal by polarisation diplexing before being transmitted to the two receiver clusters.

Table I shows the basic technical characteristics of the system.

The SST will be operated at the El Leoncito Astronomical Complex, CASLEO in the Argentinean Andes (Province of San Juan). Its high altitude of 2.500 m, the very low atmospheric water vapour content and the nearly 300 clear days per year will provide high atmospheric transparency for submm waves. Most of the days during the 6 cooler months have a total water vapour content considerably less than 2 mm (Filloy and Arnal, 1991).

The main scientific objectives of the SST project can be summarized as follows:

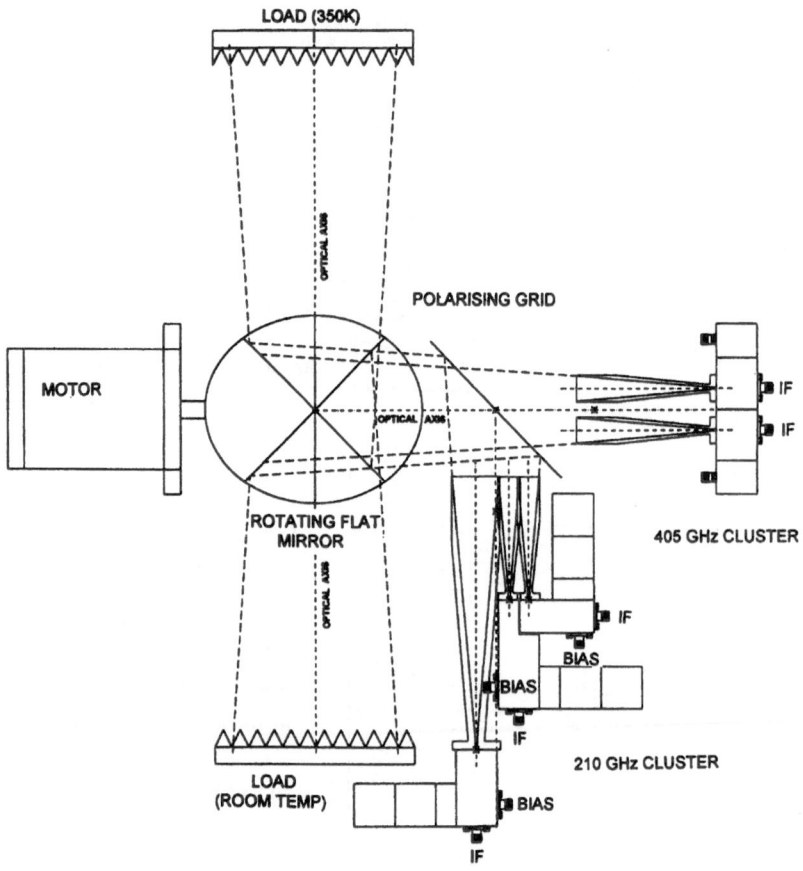

Fig. 2. The front-end with the radiometers as seen from top (four at 210 GHz and two at 405 GHz), after RPG.

1. The explosive component of solar flares with the emphasis on:
   (a) the extension of burst spectral components into the submm-wave range
   (b) the identification of fast structures in bursts with ms time resolution and a sensitivity of down to 0.1 s.f.u.
   (c) the search for the existence of a new submm-IR spectral component
   (d) the association between the emission of submm-wave bursts and complementary observations at other wavelengths, such as in the x/gamma-ray region, in optical lines and the whole radio spectrum. Further information on source structure and the emission spectrum at lower mm-wavelengths will be available from Itapetinga (48 GHz) and BIMA (86 GHz)

(e) the measurement of the position of burst centroid emission at 210 GHz with ms time resolution by using the multiple beam technique.

2. The quiet Sun, with particular attention to:
   (a) the radio evolution and importance of active regions, and
   (b) the central solar disk temperature and limb profile at mm/submm wavelengths.

Subsystem	Description	Manufacturer	Notes
Reflector	1.5-m Cassegrain, f/D=8, overall r.s.s. better than 30 μ, aluminum surface, back-structure and thermal interface ring,	University of Arizona Steward Observatory Tucson, AZ, USA	Subreflector can be focused axially
Receivers (405 GHz)	2 receivers, optimum feed-horn taper; system noise temperature < 3000 K; 2 IF-channels for solar (0.5-1.5 GHz DSB) and ClO line (390 GHz) measurements (14.3-14.7 GHz)	RPG-Radiometer Physics, Meckenheim Germany	HPBW of approx. 2 arcmin; delivery expected for March 1997
Receivers (210 GHz)	4 receivers, one with optimum taper feedhorn, three in cluster with small taper, producing beams overlapping at half power points; system temperature < 3000 K; IF 0.5-1.5 GHz, DSB	RPG-Radiometer Physics, Meckenheim Germany	HPBW of approx. 4 arcmin
Positioner	El-Az, inductosyns, 3.6 arc-sec accuracy and 12 arcsec repeatability; max. speed 2 $deg \cdot sec^{-1}$, max. accel. 2 $deg \cdot sec^{-2}$	ORBIT Advanced Technologies Netanya, Israel	Astronomical pointing calibrations may improve measured specs
Radome	2.7-m height, 3.3-m diameter, Gore-Tex membrane on metal space frames	ESSCO Concord, MA, USA	high transparency for submm waves

Table I - Principal technical characteristics of SST

3. Radio-propagation experiments in the 200 and 400 GHz bands for the study of:

(a) the atmospheric absorption as a function of meteorological parameters on the ground, time of the day and season

(b) the atmospheric irregular refraction ("seeing") by using artificial point sources (beacons) in the antenna far field and by solar limb tracking, sensitive to angular displacements smaller than one arcsecond,and

(c) the spectral line measurements of ClO at 390 GHz.

4. Additional technical benefits will be obtained by studying the system performance as:

(a) effects of antenna thermalisation on beam pattern and efficiency; positioning accuracy as a function of temperature, and

(b) measurements of antenna beam patterns by using beacon transmitters and also stronger celestial sources as the Sun,Moon and planets.

The SST project was funded by Brazil's S. Paulo State Foundation for Support of Research (FAPESP, Contract 93/3321-7), and receives partial funding from co-participants: Argentina's organizations Institute of Astronomy and Space Physics (IAFE), Buenos Aires, and CASLEO, and Switzerland's Institute of Applied Physics, IAP, of the University of Berne. The SST project is expected to become operational in 1998.

References

Costa J.E.R., Correia E., Kaufmann P., Magun A., and Herrmann R. (1995): Solar Phys. **159**, 157.

Filloy E., Arnal F.R. (1991): Poster at 21st IAU General Assembly, Buenos Aires, Argentina.

Georges C.B., Schaal R.E., Costa J.E.R., Kaufmann P., and Magun A. (1989): *50 GHz multi-beam receiver for radio astronomy*, in Proc. 2nd Int. Microwave Symp., Rio de Janeiro, p. 447.

Herrmann R., Magun A., Costa J.E.R., Correia E., and Kaufmann P. (1993): Solar Phys. **142**, 157.

Solar Astronomy and the Square Kilometer Array Interferometer

Robert Braun

Netherlands Found. for Research in Astronomy, Postbus 2, 7990AA Dwingeloo, NL

Abstract. A next generation radio observatory, the "Square Kilometer Array Interferometer", is now being designed to provide a leap in sensitivity of about two orders of magnitude over current facilities operating at frequencies between about 200 MHz and a few GHz. Such a leap in performance is motivated by a wide range of fundamental astrophysical questions which we hope to address in the first decades of the next millenium. Some of the myriad possibilities generated by this instrument are demonstrated by considering it's impact on the study of the evolution of galaxies. We then turn specifically to the possibilities offered for extending the frontiers of solar astrophysics. Timely consideration of the special constraints imposed by solar imaging should make it possible to achieve an increased sensitivity by a factor of 10^5 over current levels. "Speckle" imaging at sub-arcsecond resolution of even the quiet solar surface should then be practical over much of the frequency range. "Solar microscopy" may also become possible, in which near-field effects are employed to obtain the range as well as position of high brightness solar emission features.

Résumé. Le projet "Square Kilometer Array Interferometer" a pour but de concevoir un radio téléscope de nouvelle génération avec une surface de détection voisine du kilomètre carré, permettant d'observer dans un domaine de fréquences allant d'environ 200 MHz à quelques GHz. Un tel instrument permettra de gagner environ deux ordres de grandeur en sensibilité par rapport aux instruments actuellement en opération dans cette bande de fréquences. Ce gain considérable des performances instrumentales permettra d'apporter des réponses à un grand nombre de questions fondamentales de l' astrophysique. Nous présentons quelques unes des innombrables possibilités ouvertes par ce nouvel instrument en discutant de son impact sur l'étude de l'évolution des galaxies. Cet instrument permettra également de faire reculer les frontières de l'astrophysique solaire. En effet, les contraintes spécifiques, imposées par l'imagerie du Soleil, seront prises en compte. et un gain en sensibilité de $\sim$ 10^5 devrait être atteint. L'imagerie des tavelures, avec des résolutions spatiales inférieures à la seconde d'arc, deviendra donc possible, même pour le Soleil calme, dans la majeure partie de la bande observée. On pourra également utiliser les effets de champ proche pour detecter et localiser une grande variété de micro événements de forte brillance.

1 Introduction

Since it's invention in the 1930's, radio astronomy has been applied with great success to a wide range of astrophysical problems. Important physical insights continue to be made into the nature of both condensed and diffuse objects which lie at distances ranging from within the solar system to the recombination surface of the bubble which defines our visible universe.

A basic requirement for the continued success of our quest for greater physical insight is the need for a continuous improvement in instantaneous sensitivity. In one way or another, all of the various research directions in radio astrophysics are limited by our current instrumental sensitivities. Only by insuring the continued access to order of magnitude improvements in our capabilities can we insure a continued high rate of discovery. This statement is put into perspective by considering how our capabilities have evolved to their present level. In Figure 1 we have plotted the continuum sensitivity after one minute of integration for many radio telescopes as they came on-line. Substantial improvements to the performance from post construction up-grades to the receiver systems are also indicated in the figure for some of the facilities. An exponential improvement in system performance, over at least 6 orders of magnitude can be seen between about 1940 and 1980. Instruments like the Westerbork Synthesis Radio Telescope (WSRT) and the Very Large Array (VLA) have become available on a schedule which maintained a high rate of discovery. The Arecibo telescope stands out as a major leap in sensitivity performance at a relatively early date. This particular example serves to illustrate that the single parameter we've chosen to examine in Figure 1, the continuum sensitivity in a short integration, doesn't necessarily tell the entire story. Additional parameters, like sky coverage, spatial resolution and survey speed also play a significant role in defining the total system performance. Even so, a disturbing trend seems to have developed in the period since about 1980. There appears to be a significant saturation of performance obtained with traditional radio telescope technology.

The time may now be ripe to explore a radically different technology in the decimetric radio band; one in which the economies of mass production are applied to high performance, yet extremely low cost amplifiers and digital electronics, while the dependence on large mechanical components is minimized. Just as dinosaurs were superseded by mammals, we should consider moving to distributed networks of smaller, yet more intelligent components. Hopefully, this transition can be realized without a major catastrophe first befalling the entire field. Rather than beginning with a mass extinction, we can hope for a fruitful period of coexistence of current and next generation radio astronomy instrumentation.

The provisional name that has been given to such a next generation instrument, is the "Square Kilometer Array Interferometer", or SKAI. This name embodies the fact that on the order of a square kilometer of collecting area will be required to provide the leap in sensitivity depicted in Figure 1.

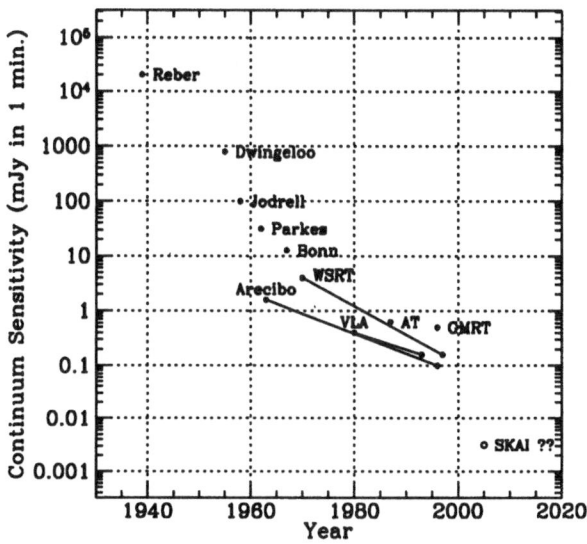

Fig. 1. The time evolution of radio telescope sensitivity. The continuum sensitivity after one minute of integration is indicated for a number of radio telescopes as they became available. Solid lines indicate up-grade paths of particular instruments.

2 Scientific Drivers for the Array

Detailed accounts of specific scientific drivers for such a high sensitivity instrument can be found in several recently published volumes (Raimond and Genee 1996, Jackson and Davis 1996). Scientific breakthroughs can be foreseen in understanding the:

• kinematics and evolution of both nearby and very distant galactic systems utilizing emission and absorption in the HI 21 cm line

• kinematics and evolution of circum-nuclear starburst systems utilizing molecular mega-maser emission

• origin and evolution of galactic disks and halos utilizing radio continuum emission

• origin of pulsar emission and its utilization as a probe of extreme states of matter

• evolution and physical properties of normal stars including the Sun

• AGN and jet phenomenon utilizing the highest possible angular resolution

Rather than reiterating those discussions here, we will briefly demonstrate the impact of a next generation radio observatory with only a few specific examples and comments related to understanding the evolution of galaxies in the next subsection. Then we will turn to issues related more specifically to solar radio astronomy.

2.1   The Evolution of Galaxies

The 21-cm line of neutral hydrogen remains the most valuable tracer of neutral gaseous mass in astrophysics. Even though neutral gas becomes predominantly molecular at high densities, the 21-cm line emission of the atomic component allows the total gaseous mass to be estimated to better than about a factor of two, even in the most extreme cases. This is in marked contrast to, for example, the luminosity of carbon monoxide emission lines originating in the molecular component. For the same total gaseous mass, the CO emission lines can vary in luminosity over a factor of about 10^4 depending on the abundance of heavy elements and the intensity of the radiation field to which they are subjected.

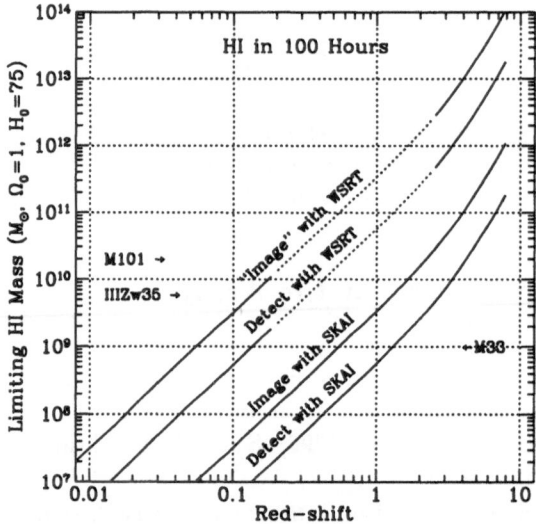

Fig. 2. We show the detection (5σ in 50 km s^{-1}) and imaging (5σ in six channels of 50 km s^{-1}) limits of atomic gas mass as a function of redshift with current and next generation instrumentation.

The current and next generation capabilities for detecting HI in emission are contrasted in Figure 2. While we must now be content to image galaxian gas masses out to red-shifts of only 0.1 to 0.2, this would become possible to red-shifts greater than 2, opening the way to truly understanding the evolution of gaseous into stellar mass.

An illustration of the enormous impact this capability would have on cosmological studies is given in Figure 3. In this figure the detection threshold of SKAI is overlaid on the predicted mass distribution from a recent simulation of structure formation in the early universe (Katz *et al.* 1996). Even the hundreds of detections indicated in this figure understate the true potential of the instrument, since the simultaneous frequency sampling would be at least 20 times deeper than that illustrated. Such large unbiased samples should allow a clear distinction to be made between competing theories for the nature of our universe.

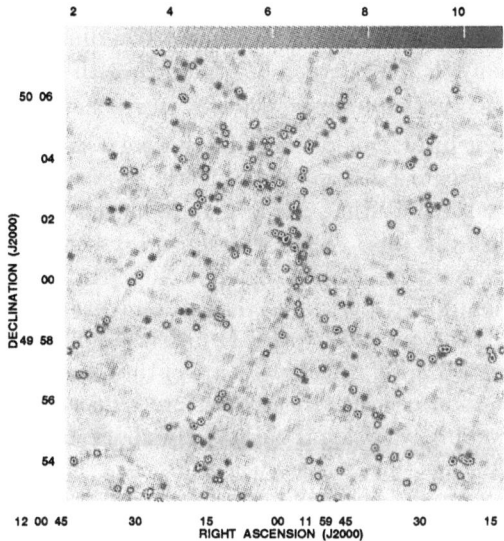

Fig. 3. Simulated HI emission at $z = 2$ with SKAI detections overlaid. The linear grey-scale indicates the predicted peak brightness of HI emission in a $22.2/(1+z)$ Mpc cube and extends from $log(M_\odot/Beam) = 1.7 - 10.8$. The single white contour at $log(M_\odot/Beam) = 9.22$ is the 5σ SKAI detection level after a 1600 hour integration.

2.2 Solar Radio Astronomy

One of the most versatile instruments for solar imaging at radio frequencies during the past decade has been the Very Large Array (VLA). Broad frequency coverage in discrete bands centered between 325 MHz and 22 GHz (and most recently 43 GHz) and a variable antenna configuration with maximum baselines between 300 m and 30 km have provided the opportunity for imaging a wide variety of solar phenomenon.

However, in a number of respects the VLA has proven less than optimal for solar imaging problems. For a variety of reasons the dynamic range of solar images has remained modest; only a few times 100:1 in even the best of cases. Both calibration quality and instantaneous spatial frequency coverage play a role in determining the dynamic range. The logarithmically spaced "wye" configuration of the VLA, while clearly superior to a linear array in terms of instantaneous imaging quality, provides sampling which is simply too limited for a complex source that fills the entire telescope beam.

Improving the dynamic range in a next generation radio facility must be addressed with a two-pronged approach: in the first instance, the special calibration requirements of solar observing must be taken along in the design phase of the instrument and secondly, much better instantaneous spatial frequency coverage must be insured via a fully two-dimensional aperture distribution of as many elements as can be obtained, with a minimum of perhaps 30, but a preference for as many as 100.

A second major limitation of the VLA for solar astronomy has not generally even been recognized as such due to the presence of the first. This is the very limited sensitivity that can now be achieved in a solar observation. Since the solar disk fills most or all of the primary beam of a 25 m diameter paraboloid observing at frequencies of a few 100 MHz or higher, the system temperature is dominated by the brightness temperature of the solar disk. For example, while the nominal system temperature of the VLA in the 20 cm band is 35 K, the VLA antenna temperature of the quiet sun at this frequency is 42000 K. The system sensitivity is therefore reduced by a factor of more than 1000 over nominal values, providing only about 1 Jy per beam rms in a 30 second integration.

While this may be ample sensitivity for the imaging of active regions at modest resolution, it is woefully inadequate for direct and speckle imaging of faint emission features at sub-arcsecond resolution. Bastian (1994) points out the likely contributions of turbulent scattering in the solar corona to the time averaged source size of emission features. Scattering sizes of order 20 arcsec seem to be indicated for an observing wavelength of 20 cm. "Freezing" the scattering into speckles of the intrinsic source size is likely to demand a spectral resolution of 100 Hz and time resolution of 70 msec at this observing wavelength. Unfortunately, as Bastian points out, the VLA brightness sensitivity on relevant baselines with these parameters is a depressing $2\text{--}4 \times 10^9$ K, while the source brightness is likely to be only a few times 10^6 K for quiet sun surface features and a few times 10^9 K for incoherent bursts. It is this discrepancy in sensitivity which is responsible for the coarse limiting resolution of current generation instruments for solar imaging at low frequencies.

Improving the instantaneous sensitivity of solar observations also requires a two-pronged approach. The factor of 1000 sensitivity loss noted above for 25 m class paraboloids can be circumvented by employing a much smaller basic element size. For example, the antenna temperature due to the sun seen by a 1 m element at λ 20 cm is only 3 K, adding only negligibly to the system temperature. Large numbers of such small elements can be combined with appropriate complex weights so that the effective antenna temperature is due almost exclusively to the *correlated* power. Further gains in sensitivity come directly from increasing the total collecting area. The 100-fold increase in collecting area being considered for SKAI then provides a total enhancement in sensitivity of 10^5 over the VLA for solar observing. Suddenly, even the quiet sun would be accessible to sub-arcsecond imaging at decimetric wavelengths.

Another possibility that has not yet received serious consideration is that of undertaking earth-based microscopy of the Sun. The Sun actually resides in the near-field with respect to moderately long baselines and relatively high radio frequencies. Since there is a slightly different wavefront curvature for the spherical wavefronts originating at different distances along the line-of-sight, it becomes possible to assign a *radial* position to particular emission features as well as the spatial coordinates. A phase precision, of $\Delta\phi$ radians, on a baseline of length B, provides a radial resolution, ΔR at the solar distance, $D_\odot$, of

$$\Delta R \approx \frac{4c\Delta\phi D_\odot^2}{2\pi\nu B^2} \qquad (1)$$

or

$$\Delta R \approx 0.12 \left(\frac{\Delta\phi}{0.1\ deg}\right) \left(\frac{\nu}{1\ GHz}\right)^{-1} \left(\frac{B}{300\ km}\right)^{-2} \quad R_\odot. \qquad (2)$$

Phase precisions of 0.1 degree are already sufficient to provide a radial resolution of 0.1 $R_\odot$ at 1.5 GHz on a 300 km baseline. Of course, obtaining the necessary phase precision on a 300 km baseline assumes the presence of sufficient correlated power at that frequency. As just discussed above, the small correlation bandwidth and integration time at 1.5 GHz lead to a brightness sensitivity per 300 km baseline of about $10^{7.5}$ K. Such observations might then only be conceivable for events with rather high brightness temperatures, in excess of about 10^9 K. Solar microscopy might be more easily accomplished at higher frequencies, near 10 GHz say, where larger correlation bandwidths and integration times can be employed.

3   Technical Specifications

Consideration of the many varied scientific drivers suggests the following basic technical specifications for the instrument:
• a primary frequency range of about 200 – 2000 MHz, to address the (redshifted) spectral lines of HI and OH, with a strong desire for yet higher frequency coverage particularly for VLBI and stellar applications,
• a total collecting area of about 1 km^2, to achieve the desired sensitivity when employed with nearly sky-limited system temperatures,
• distribution over at least 32 elements, to achieve a sufficiently clean instantaneous synthesized beam and permit adequate modeling of time variable (interfering) sources
• a preference for forming these 32 elements from many smaller ones, so as to allow adaptive beam formation to excise interfering sources,

• better than about 1 Kelvin of brightness sensitivity for spectral line applications, implying a maximum array size of about 50 km

• spatial resolution of 0.1 to 1 arcsec for continuum applications, implying an array size of about 300 km

One way to satisfy the conflicting demands of spectral line and continuum applications apparent above is to place a large fraction (say 80 percent) of the collecting area of the instrument within a region of about 50 km and to distribute the remaining fraction over a region of about 300 km diameter. In this way, the applications most limited by low surface brightnesses would retain most of their sensitivity, while other applications could utilize angular resolutions as high as about one tenth of an arcsecond.

A very schematic representation of what the SKAI might look like is given in Figure 4. The shaded circles indicate the position of the unit telescopes of the array. Note that the size of each circle has been greatly exaggerated to allow it to be seen on the scale of the illustration. A rather dense ring-like concentration of the telescopes over a region of about 50 km extent determines the beam size for which the brightness sensitivity is optimized (about 1 arcsecond at a frequency of 1420 MHz), while the additional elements, distributed over a 300 km extent, would make sub-arcsecond imaging possible over the entire frequency range of 200 – 2000 MHz. The thick annular distribution of the elements within 50 km is chosen to provide an approximately Gaussian naturally weighted synthesized beam. Although not indicated accurately in the Figure, care would obviously be taken to insure good sensitivity to short projected baselines and the elimination of any "holes" in the spatial frequency coverage.

Other authors (Swarup, 1996) have suggested that it might be desirable to place as much as 50% of the total collecting area in one contiguous region, as has been done with the GMRT. Such a configuration would provide enhanced sensitivity to spatial scales of about 1 arcmin, while reducing the sensitivity to spatial scales of several arcsec. Trade-offs of this type will need to be made on the basis of our accumulated knowledge during the final design phase. In the mean time, the GMRT will have ample opportunity to demonstrate the utility of the course they have chosen to follow.

4   Telescope Concepts

Several possible element concepts for the SKAI are illustrated in Figure 5. At the heart of these concepts is a much greater reliance than ever before on mass produced and highly integrated receiver systems together with much more extensive digital electronics for beam formation. In the top panel we depict one conceivable extreme in a continuous range of possibilities. In this case the wavefront is detected by individual active elements comparable to a wavelength in size. Each of these is amplified, digitized and combined with the others to form an electronically scan-able beam (or beams) with no mov-

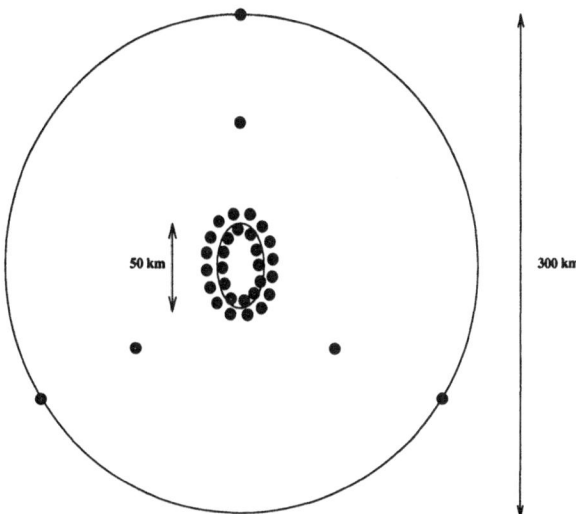

Fig. 4. Schematic configuration of SKAI. Note that the unit telescopes are not depicted to scale, nor are the telescope locations accurately specified.

ing parts whatsoever. The challenge in this case lies in achieving extremely low component and data distribution costs since literally millions of active elements will be required. In the center panel, some degree of field concentration is first achieved with the use of small reflectors before amplification, digitization and beam formation. In this case, active element number is reduced to some thousands and greater sky coverage at high sensitivity is also realized, although at the expense of the mechanical complexity of the drive and tracking system.

The adaptive beam formation technology which underlies these element concepts is extremely attractive for a number of reasons. Real-time beam formation with at least thousands if not millions of active elements provides a comparable number of degrees of freedom for tailoring the beam in a desired way. The basic properties of high gain in some direction and low side-lobe levels elsewhere are fairly obvious and traditional requirements. An additional possibility, which hasn't yet been applied in radio astronomy, is that of placing response *minima* in other desired directions, such as those of interfering sources. In addition, the way is naturally opened to exploit multiple observing beams on the sky to enhance the astronomical power of the instrument many-fold. These might be used to provide simultaneous instrumental calibration, support multiple, fully independent observing programs or enlarge the instantaneous field-of-view for wide-field applications. Finally, the great potential of adaptive beam formation has led to a strong commercial interest

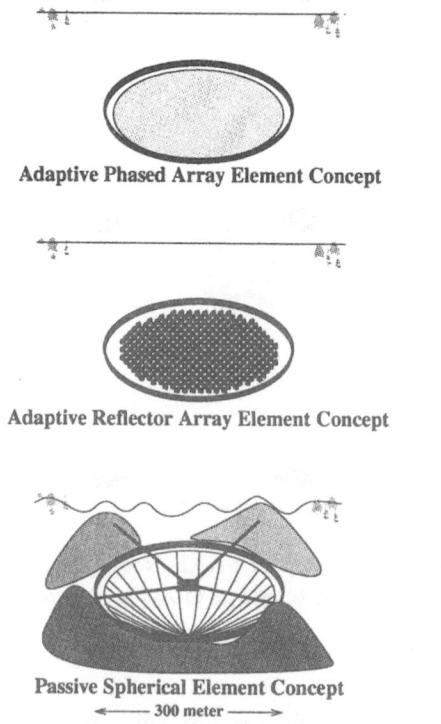

Adaptive Phased Array Element Concept

Adaptive Reflector Array Element Concept

Passive Spherical Element Concept
◄──── 300 meter ────►

Fig. 5. Possible adaptive element concepts for the SKAI. Fixed spherical and adaptive parabolic reflector concepts are also being studied.

in this technology. This has opened the way to collaborative R&D efforts which are now beginning to take shape.

During the interval 1995–2000, a concerted effort at R&D for SKAI will be undertaken both within the NFRA and at collaborating institutes. The concepts depicted in Figure 5, as well as others utilizing large fixed (like Arecibo) or adaptive reflectors will be worked out in sufficient detail to allow realistic cost estimates to be made. Proto-typing of cost effective technologies as an extension to the WSRT array is planned for the period 2001–2005. Assuming the successful completion of both technical preparations and funding arrangements, construction of the instrument is envisioned for the period 2005–2010. In this way we can look forward to opening the door to a new era of discovery.

References

Bastian T.S. (1994) Angular Scattering of Solar Radio Emission by Coronal Turbulence ApJ **426**, 774

Jackson N., Davis R. (1996) *High Sensitivity Radio Astronomy* (Cambridge Univ. Press, Cambridge)

Katz N., Weinberg D.H., Hernquist L., Miralda-Escudé J. (1996) Damped Lyman-Alpha and Lyman Limit Absorbers in the Cold Dark Matter Model ApJ **457**, L57

Raimond E., Genee R. (1996) *The Westerbork Observatory: Ongoing Adventure in Radio Astronomy* (Kluwer, Dordrecht)

Swarup G. (1996) *Array Configuration for a Radio Telescope of One Million Square Meter Area* (URSI GA Abstracts)

Prospects for the Solar Radio Telescope

Timothy S. Bastian[1] and Dale E. Gary[2]

[1] NRAO, P.O. Box 'O', Socorro, NM 87801, USA
[2] Caltech, Solar Astronomy, Pasadena, CA 91125 USA

Abstract. The Solar Radio Telescope (SRT) is an instrument concept for a powerful solar-dedicated radio telescope. As presently conceived. it would combine a high-resolution imaging capability ($2''$ at 20 GHz) with a broadband spectroscopic capability (0.3–26 GHz). In other words, the SRT would perform broadband imaging spectroscopy on a wide range of quiet- and active-Sun phenomena. On 17–20 April, 1995, a workshop was held in San Juan Capistrano, California. The purpose of the workshop, which was attended by more than 40 scientists from the US and around the world, was to discuss the science that could be done with a solar-dedicated radio synthesis telescope, and to discuss the design constraints imposed by the science envisioned. Special attention was also given to nighttime uses for the instrument. We summarize the "strawman" concept for the instrument here.

Résumé. Le "Solar Radio Telescope" (SRT) est le concept d'un puissant radio télescope dédié à l'observation du Soleil. L'idée directrice est de combiner imagerie à haute résolution spatiale ($2''$ à 20 GHz) et spectrographie dans un domaine de fréquences étendu (0.3–26 GHz). Ceci premettra d'obtenir, dans une large bande spectrale, des images d'une grande variété de phénomènes qui se produisent dans l'atmosphère solaire qu'elle soit calme ou active. Un atelier de travail international, qui a réuni plus de 40 scientifiques américains et de nombreuses autres nationalités, s'est tenu à San Juan Capistrano, Californie, du 17 au 20 avril 1995. Cette réunion a permis de dégager les objectifs scientifiques que permettrait d'atteindre un grand radio télescope à synthèse dédié au Soleil et de discuter des containtes instrumentales qu'imposent ces objectifs. On y a également souligné l'intérêt d'un tel instrument pour des observations non solaires pendant la nuit. Dans cet article nous résumons les grandes lignes du concept expérimental et des objectifs scientifiques.

1 Introduction

Large aperture synthesis telescopes operating at decimetric and centimetric wavelengths (e.g., the WSRT and the VLA) have contributed to an improved understanding of a variety of solar phenomena over the past 15-20 years. However, since these instruments are optimized for observing cosmic sources, and are in any case only available to solar observers for a relatively small amount of time each year, progress in exploiting radio diagnostics of physical processes on the Sun in this wavelength regime has been slow. Recently, several solar-dedicated telescopes have been upgraded or constructed. These include the Nançay radioheliograph (Kerdraon and Delouis 1996), the OVRO Solar

Array (Gary and Hurford 1994), and the Nobeyama radioheliograph (Nakajima et al. 1993) – instruments which are producing valuable interferometric data on a routine basis. The Nançay radioheliograph operates in decimetric and decametric bands. The NRO radioheliograph operates at 17 and 34 GHz (Takano et al. 1996). The OVRO Solar Array, while frequency agile between 1–18 GHz, has only 5 antenna elements and, therefore, has a limited imaging capability. Each of these instruments is located at widely different longitudes which largely precludes joint observations of transient phenomena. In order to obtain *comprehensive* observations over the decimetric and microwave bands, *a solar-dedicated broadband imaging spectrometer optimized for solar imaging is needed.* At a workshop in San Juan Capistrano, California, in April 1995, the scientific justification for such an instrument was considered and, in light of scientific concerns, a strawman instrument concept was developed. Details may be found in a document edited by Gary & Bastian (1996).

The instrument design calls for a frequency-agile synthesis telescope capable of performing broadband imaging spectroscopy and polarization measurements on timescales $\lesssim 1$ s. It will be composed of 1–3 large (~ 25 m) elements and 20–40 small (2 m) elements outfitted with feeds and receivers operating between 300 MHz and 26 GHz. Furthermore, the large elements may be outfitted with a backend for high-time and high-frequency resolution decimetric spectroscopy.

The SRT will be the *only* instrument, space- or groundbased, capable of producing high-resolution images of thermal and nonthermal phenomena over the decimetric/centimetric wavelength range. Phenomena for which the array will potentially have a major impact on our understanding include solar flares, erupting filaments and prominences, and coronal mass ejections. The array will also have excellent sensitivity for quiet Sun studies, including high-resolution imaging spectroscopy of the solar chromosphere, coronal magnetography in solar active regions, and high-resolution studies of the formation and evolution of filaments and prominences.

While solar-dedicated by day, the instrument will possess unique capabilities for observing cosmic sources at night and could be used on a "target-of-opportunity" basis during the day. The large elements, needed for calibration of the small antennas, can be optimized for cosmic work, with broadband spectral capabilities and high sensitivity. Such a system would offer opportunities for longterm monitoring projects, broadband spectroscopy of transient sources (e.g., novae, active stars, scintillating sources), and a rapid response to new sources of interest (e.g., super-novae, gamma-ray bursts).

2   The Strawman Concept

The primary goal is to design and construct an instrument which fully exploits solar microwave emission as a diagnostic of physical processes on the Sun. To this end, a number of instrumental requirements have been identified:

1. *Imaging capability:* The sources of radio emission on the Sun must be imaged with high dynamic range, fidelity, and angular resolution with good sensitivity to both compact and extended sources of emission. A dynamic range of order 1000:1 and angular resolution of 2″ at a frequency of 20 GHz are considered reasonable goals.
2. *Support of broadband spectroscopy:* The brightness temperature spectrum as a function of position is required for both decimetric and microwave phenomena. Hence, spectroscopic coverage over the frequency range of 300 MHz to $\gtrsim$ 26.5 GHz with a spectral resolution of $\sim$ 5% is needed.
3. *Support of polarimetry:* At least partial support of polarimetry is needed; i.e., measurements of right- and left-circularly polarized radiation (see §3 below).
4. *High time resolution:* Brightness temperature spectra must be acquired at a rate which resolves the timescale on which the phenomenon of interest evolves. The most demanding requirement is imposed by the impulsive phase of flares, which will require a time resolution of < 1 sec.
5. *Large field of view:* In the interest of maximizing observing efficiency and in matching the capabilities of many full disk spectrographs and imagers, a full disk imaging capability is desired at most frequencies.
6. *Good absolute positional accuracy:* Instruments in most wavelength bands now possess an angular resolution range from less than 1 arcsec to several arcsec. Quantitative cross-comparisons between various wavelength regimes will require absolute source positions.

These requirements lead us to propose the following *strawman design*. First, the requirement of high angular resolution imaging suggests an instrument which employs Fourier synthesis imaging using an interferometric array of antennas. The requirements of high dynamic range and image fidelity, and of high sensitivity to a wide range of angular scales suggest many antenna elements are needed – perhaps 20-40 antennas. An angular resolution of 2″ at a frequency of 20 GHz requires a maximum antenna baseline of $\approx$ 1.5 km while good sensitivity to extended emission will require adequate numbers of short antenna baselines. It is anticipated that frequency synthesis techniques will also be exploited (e.g., Bastian 1989), perhaps in the form of spatial/spectral image reconstruction (Komm, Hurford & Gary 1996).

Second, the requirement of full disk imaging at most frequencies suggests the use of small antenna elements. The strawman design calls for apertures of 2 m. On the other hand, the need for good absolute calibration implies a need for astronomical calibration; i.e., sufficient sensitivity is needed to compare the source position with that of known cosmic reference sources. This, in turn, implies a need for large, sensitive antennas. Hence, in addition to many small antennas, the strawman design calls for at least one large antenna. The construction of one or more new 25 m antennas would be prohibitively expensive. An affordable alternative is to refurbish existing antennas—e.g., the two 27 m antennas at Owens Valley, or the three 25 m antennas at Green Bank. These will be used to calibrate the instrument against cosmic standards, to perform high resolution decimetric spectroscopy (§3), and to perform nighttime observing (§4)

Third, the requirement of broadband spectroscopy in the decimetric and microwave bands suggests a frequency-agile design. The technology for the detection, transmission and amplification of broadband microwave signals is mature, and off-the-shelf components are available at low cost in several bands of interest (e.g., 1–8 GHz, 8-18 GHz, 18–26.5 GHz, etc.). A match to commercially available microwave bands is therefore desirable, and an initial core operating frequency of 1–26.5 GHz is called for in the strawman design. Both high- and low-frequency extensions are feasible: to 300 MHz at low frequencies, and up to 40 GHz and/or the 80–115 GHz band at high frequencies.

Finally, the requirement for high time resolution may appear problematic in view of the large number of baselines and the broadband frequency coverage required – to process a bandwidth of 1–26.5 GHz for an array of 40 antennas (or 780 baselines) instantaneously would require a very large correlator. However, the entire spectrum need not be sampled instantaneously and the frequency resolution and rate at which spectra are acquired depends on the phenomenon of interest. The high flux levels from the Sun allow an extremely fast sampling rate (~ 10 msec) with good signal to noise. Thus, the strawman design assumes that 500 MHz wide sections of the total bandwidth will be sampled sequentially; i.e., that frequency multiplexing will be performed. With the wide instantaneous IF bandwidths (500 MHz) now available, the entire 1-26 GHz band would be covered with ~ 50 samples in <1 sec. In other words, during flares, a set of 50 maps would be produced each ~ 1 sec. It would be wise to design the IF/LO system and correlator with sufficient flexibility that the specific frequencies sampled, their number, and the rate at which they are sampled, are tunable parameters. In this way the instrument can be used most effectively for imaging spectroscopy of a wide variety of transient phenomena.

3   Daytime Observing

The primary goal of the SRT is to support solar-dedicated broadband imaging spectroscopy of the Sun. When the SRT is brought into operation, it will have a profound and broad impact on solar physics. In terms of solar activity (flares, eruptive events, and other particle acceleration events), the SRT will address the following:

1. *Location and dynamics of coronal flare sources:* The SRT will give high-dynamic-range, full-disk images of the high-energy nonthermal and superthermal component of flares with a quality comparable to that available in soft X-rays from *Yohkoh*, with a timescale of < 1 s. The broad

frequency range covered by SRT guarantees that the source distribution will be imaged over the full range of relevant coronal and particle parameters.

2. *Spectral diagnostics of the electron distribution function:* At each point in these full-disk images, a full brightness temperature and polarization spectrum will be obtained. These spectra have shapes that depend in characteristic ways on the electron distribution function and other plasma parameters, including magnetic field strength. Detailed quantitative values for these parameters can be derived or modeled for both thermal and nonthermal sources.

3. *Particle acceleration and eruptive phenomena:* Radio emission is sensitive to both cool chromospheric material (through free-free emission) and nonthermal or hot thermal particles (through gyrosynchrotron emission). When particle acceleration occurs in conjunction with eruptive phenomena such as filaments and CMEs, the SRT will obtain detailed, simultaneous images and spectra of both phenomena, giving accurate timing and spatial correspondence as well as spectral diagnostics. Simultaneous observations of dm- and m-λ radio bursts could also be done when the frequency range is extended to 300 MHz.

4. *Comparisons with SXR, HXR, and γ-ray emissions:* The same electrons that produce these emissions also produce radio emission, and as such there is a high degree of complementarity between radio and X-ray emissions. The X-ray observations give temperature and emission measure diagnostics, and in the case of hard X-rays can show the presence of nonthermal particles. Radio observations give similar but complementary information, so comparisons between the two give more complete information than either can do alone.

5. *Studies of buildup and preflare phenomena:* Changes in the corona that precede and perhaps lead to the main energy release event in a flare can be studied in exquisite detail in the radio regime. The sensitivity of radio emission to small numbers of nonthermal electrons, or to small changes in thermal emission, would allow spectral imaging studies of phenomena such as preheating and enhanced particle acceleration, and even slight changes in direction and magnitude of magnetic fields (see below).

6. *Direct imaging of CMEs:* Recently, Bastian and Gary (1997) explored the question of whether an instrument like the proposed SRT could detect coronal mass ejections (CMEs). They show that, provided differential detection techniques are employed, the thermal bremsstrahlung emission from CMEs could indeed be detected. One of the important strengths of the SRT, however, is that owing to its frequency agility it could provide a comprehensive observational picture of CMEs and associated phenomena over the entire decimetric and microwave band of frequencies. For example, in addition to imaging the radio counterpart of a CME through the thermal bremsstrahlung radiation it emits, the SRT could also ob-

serve the microwave emission from an associated flare and, in some cases, the high frequency tail of associated bursts of type II or type IV. As such, the SRT will serve as a powerful tool for disentangling the complex relationship between CMEs, flares, filament eruptions, and radio bursts.

In studies of quiet Sun and quiescent structure of the chromosphere, transition region, and corona, the SRT would address the following:

1. *Structure of solar atmosphere:* The SRT will obtain spatially and temporally resolved measurements of the Sun's outer atmosphere as a function of optical depth. Tuning to different frequencies can be likened to scanning in optical depth, and hence height in the atmosphere. Intensity and polarization information as a function of optical depth can in principle be inverted to determine the physical properties of the atmosphere (temperature, density, and magnetic field strength) as a function of height.

2. *Birth, evolution and death of active regions:* The solar-dedicated SRT will obtain continuous, daily coverage (8-12 h/day) of active region evolution. Data products include coronal magnetic field maps obtained through inversion of the gyroresonance spectrum in sunspots and plage areas of active regions. The ability to measure coronal magnetic fields and watch the growth and decay of sunspots at the coronal level has application to coronal heating, coronal currents, and a number of other fundamental questions about the origin of solar activity.

3. *Evolution of filaments and streamer structure:* The imaging and spectral diagnostics of the SRT will also apply to filaments, filament channels, filament cavities, and prominences. Seen through their free-free absorption on the disk, or free-free emission off the limb, quantitative estimates of temperature and emission measure can be made throughout the cooler parts of the structure. Changes in the hotter parts of the structure may also be visible using differential techniques as discussed for the case of CMEs in Bastian and Gary (1997).

4. *Evolution of coronal holes:* In the quiet Sun outside of active regions, the lower atmosphere of the Sun can be observed by SRT. Coronal holes are easily detected and could be followed at lower frequencies through the solar cycle, while their underlying chromospheric structure could be deduced through inversion or analysis of the free-free emission spectrum at higher frequencies. This structure can be resolved into network and non-network components, and the slight dependence of free-free emission on magnetic field strength may allow network magnetic structure to be studied.

Operationally, the large antennas elements would be used to calibrate the small elements before and after the observing day. The large elements would then be used for cosmic programs (see below). However, a low-frequency option discussed at San Juan Capistrano was to use the large elements to perform high spectral and temporal resolution spectroscopy at decimetric wavelengths when targets of interest are present on the solar disk. The high resolution spectroscopy would provide an important complement to the decimetric imaging provided by the array.

4   Nightime Observing

The small antennas elements of the SRT will have insufficient sensitivity to be useful for observing most cosmic sources. However, the sensitivity and frequency agility of the 25 m antennas would allow the SRT to serve the "nightime" community in a unique capacity. The most optimistic case will include 3 large antennas. If we assume that the system temperature of each large antenna will be comparable to that of a present-day VLA antenna, the sensitivity of the trio of antennas would be 1/9 that of the VLA. If the SRT were to operate in an observing mode in which IF's of 500 MHz bandwidth were obtained simultaneously in each polarization, then the improvement in sensitivity compared to the VLA (2 IF's of 50 MHz bandwidth in each polarization), would be $10^{0.5}$, bringing it to more than 1/3 the VLA's sensitivity. Given that the SRT correlator will be designed to process many baselines simultaneously, part of that processing power may be brought to bear to process much more than 500 MHz bandwidth. In fact, for certain detection or monitoring experiments, it may be desirable to process $\sim$ 10 GHz of bandwidth. In such cases, the sensitivity of the SRT would exceed that of the present-day VLA. An important programmatic factor, therefore, is that the correlator required for solar observations would operate in one of two modes. During solar observations it would operate with only one or a few IF bands but with many antennas. During non-solar, nighttime observations it would operate on many IF bands at once but with only 3 antennas. Such a trade-off would make maximum use of the correlator. Low noise, cryogenic amplifiers would be required only on the 3 large dishes.

For imaging observations such an instrument would be at a disadvantage to, say, the VLA. However, the 3-element interferometer comprised of the three large dishes, coupled with frequency synthesis techniques, will allow source localization if not imaging, and will be far less susceptible to source confusion than a single-dish with similar spectral capabilities. And such a configuration would offer unique opportunities to the astrophysical community at large. First, it would be the only instrument capable of performing broadband spectroscopy on non-solar sources. Second, it would be available for at least 12 hrs a day for this purpose, and for a large fraction of daylight hours as well.

In addition to the above characteristics, we feel it is necessary to emphasize that, operationally, the SRT should: i) have the ability to respond more rapidly than existing radio facilities to "targets of opportunity." Researchers responding to the appearance of certain important classes of cosmic sources, such as γ-ray bursts, X-ray transients, novae, and supernovae, that often have radio counterparts, might request a repointing of the SRT within a very short time (minutes, hours, days) of their detection at other wavelengths; ii) support long-term monitoring and concentrated, campaigns of moderate-duration on selected cosmic sources; for example, as a part of co-ordinated, multiwavelength observing campaigns. By doing so, the SRT could fill a observational niche which has not been filled to date.

Examples of the kinds of observations the SRT could perform include:

- Monitoring the radio flux and spectral variability of quasars, AGN, high-redshift galaxies, etc.
- Searching for the radio counterparts, and obtaining radio spectra if successful, of γ-ray bursts and soft γ-ray repeaters.
- Radio studies of transient galactic sources (e.g., novae, X-ray transients, etc), X-ray binaries, super-luminal sources, and pulsars.
- Studies of interstellar scattering/scintillation of radio waves, and flux drop-outs due to small-scale spatial structure of the ISM.
- Studies of nonthermal activity in a variety of radio-emitting stars such as RS CVn and Algol binaries, T Tauri and other pre-main sequence stars, dMe flare stars, and young active stars of solar mass like AB Doradus.

In summary, the SRT offers unique capabilities to the nightime community as well as several programmatic and operational advantages over existing radio facilities for a wide range of astronomical programs.

5   Prospects for the SRT

At present, the SRT is a conceptual instrument. To make it a reality in this era of fiscal restraint requires that a number of steps be taken. First, the instrument requires broad community support. The solar radio community in the United States is small. The instrument must, at the very least, be supported by the high-energy solar physics community, and by the radio astronomical community at large. International support and/or participation in the project can only improve its prospects. Second, possible funding sources must be identified and informed about the instrument. These include the NSF, NASA, the U.S. Air Force, the Naval Research Lab, the U.S. Naval Observatory, and NOAA. Third, the astronomical community's need for new instrumentation in the United States has been prioritized through the mechanism of decade reviews. An endorsement of the SRT concept is an important goal. Finally, a detailed design study of the SRT must now proceed. This would include a site review, configuration studies, antenna design studies, studies of options for the data transmission system, the correlator, data calibration, imaging, and archiving.

Finally, it would be advantageous to design the instrument to be flexible not only scientifically, but also extendable in its construction and deployment. In this way the project could proceed incrementally and could react to shifting funding constraints.

Acknowledgement We thank the CESRA meeting organizers for the opportunity to present this talk. We are indebted to the attendees of the SRT workshop for their contributions, which led to many of the ideas expressed in this paper.

References

Bastian T.S., Gary D.E. (1997): JGR, submitted.
Gary D. E., Hurford G. J. (1994): ApJ **420**, 903
Gary D. E., Bastian T. S. (1996): The Solar Radio Telescope, Proc. San Juan Capistrano Workshop, April 1995
Kerdraon A., Delouis J.-M. (1996): this volume.
Nakajima, H. et al. (1994): Proc IEEE, **872**, 705
Takano T. et al. (1996): this volume